Communications in Computer and Information Science 2486

Series Editors

Rationale
The CCIS series is devoted to the publication of proceedings of computer science conferences. Its aim is to efficiently disseminate original research results in informatics in printed and electronic form. While the focus is on publication of peer-reviewed full papers presenting mature work, inclusion of reviewed short papers reporting on work in progress is welcome, too. Besides globally relevant meetings with internationally representative program committees guaranteeing a strict peer-reviewing and paper selection process, conferences run by societies or of high regional or national relevance are also considered for publication.

Topics
The topical scope of CCIS spans the entire spectrum of informatics ranging from foundational topics in the theory of computing to information and communications science and technology and a broad variety of interdisciplinary application fields.

Information for Volume Editors and Authors
Publication in CCIS is free of charge. No royalties are paid, however, we offer registered conference participants temporary free access to the online version of the conference proceedings on SpringerLink (http://link.springer.com) by means of an http referrer from the conference website and/or a number of complimentary printed copies, as specified in the official acceptance email of the event.

CCIS proceedings can be published in time for distribution at conferences or as post-proceedings, and delivered in the form of printed books and/or electronically as USBs and/or e-content licenses for accessing proceedings at SpringerLink. Furthermore, CCIS proceedings are included in the CCIS electronic book series hosted in the SpringerLink digital library at http://link.springer.com/bookseries/7899. Conferences publishing in CCIS are allowed to use Online Conference Service (OCS) for managing the whole proceedings lifecycle (from submission and reviewing to preparing for publication) free of charge.

Publication process
The language of publication is exclusively English. Authors publishing in CCIS have to sign the Springer CCIS copyright transfer form, however, they are free to use their material published in CCIS for substantially changed, more elaborate subsequent publications elsewhere. For the preparation of the camera-ready papers/files, authors have to strictly adhere to the Springer CCIS Authors' Instructions and are strongly encouraged to use the CCIS LaTeX style files or templates.

Abstracting/Indexing
CCIS is abstracted/indexed in DBLP, Google Scholar, EI-Compendex, Mathematical Reviews, SCImago, Scopus. CCIS volumes are also submitted for the inclusion in ISI Proceedings.

How to start
To start the evaluation of your proposal for inclusion in the CCIS series, please send an e-mail to ccis@springer.com

Dana Simian · Laura Florentina Stoica
Editors

Modelling and Development of Intelligent Systems

9th International Conference, MDIS 2024
Sibiu, Romania, October 17–19, 2024
Revised Selected Papers

Springer

Editors
Dana Simian
Lucian Blaga University of Sibiu
Sibiu, Romania

Laura Florentina Stoica
Lucian Blaga University of Sibiu
Sibiu, Romania

ISSN 1865-0929 ISSN 1865-0937 (electronic)
Communications in Computer and Information Science
ISBN 978-3-031-87385-0 ISBN 978-3-031-87386-7 (eBook)
https://doi.org/10.1007/978-3-031-87386-7

This Springer imprint is published by the registered company Springer Nature Switzerland AG
The registered company address is: Gewerbestrasse 11, 6330 Cham, Switzerland

Preface

This volume features selected, peer-reviewed papers presented at the 9th International Conference on Modelling and Development of Intelligent Systems (MDIS 2024), held from October 17 to 19, 2024, in Sibiu, Romania. The event was organized by the Research Center in Informatics and Information Technology and the Department of Mathematics and Informatics at the Faculty of Sciences, Lucian Blaga University of Sibiu, Romania. Although primarily designed as an on-site conference, a limited number of participants were permitted to attend online under specific, well-justified circumstances.

The MDIS conference series addresses topics in the modeling and development of intelligent systems, a field of significant contemporary interest. It serves as an international platform for exchanging ideas related to artificial intelligence and its applications. The 2024 edition furthered the tradition of fostering dialogue and collaboration among scientists, researchers, academics, IT professionals, and students by providing an opportunity to present and discuss their original work. Contributions ranged from theoretical and conceptual advancements to practical applications leveraging advanced technologies.

Key areas of focus for MDIS 2024 included, but were not limited to, Artificial Intelligence, Machine Learning, Data Mining, Evolutionary Computing, Nature-Inspired Algorithms, Theory of Randomized Search Heuristics, Metaheuristics and Applications, Modeling and Simulation, Mathematical Optimization, Multi-agent Systems, Intelligent Systems for Decision Support, Modeling and Development of Dynamic Systems, Security and Privacy, Cryptography, Information Retrieval, Information Systems, Software Development Methods, Search-Based Software Engineering, and Mathematical Models for Intelligent Systems.

This 9th edition welcomed participants from nine countries across three continents: Argentina, Bulgaria, Germany, Ireland, Iceland, Romania, Serbia, Slovakia, and the USA. The conference featured four distinguished keynote speakers whose presentations underscored the transformative potential of artificial intelligence across diverse domains. The keynote speakers and their respective topics were:

- Stefka Fidanova: Ant Colony Optimization and Applications
- Magda Gregorová: Deep Generative Modelling – Beyond Language and Videos
- Andrei Paun: Computing with Neuron-Like Devices: SNP Systems
- Milan Tuba: Intelligent Systems for Digital Images

All submitted papers underwent a rigorous single-blind peer-review process, with each paper evaluated by at least three independent reviewers selected based on their expertise and qualifications. This volume includes 16 carefully selected papers from a total of 44 submissions, resulting in a publication acceptance rate of 38%.

The selected papers are organized into five thematic sections: Machine Learning, Modeling and Optimization of Dynamic Systems, Intelligent Systems for Decision Support, Evolutionary Computing and Metaheuristics, and Security and Cryptography in Dynamic Systems.

We extend our gratitude to all participants for their insightful presentations and engaging discussions, as well as to the scientific committee members and external reviewers for their invaluable contributions to the academic rigor and overall success of the conference. Their efforts were instrumental in ensuring the quality of this proceedings volume.

January 2025

Dana Simian
Laura Florentina Stoica

Organization

General Chair

Dana Simian	Lucian Blaga University of Sibiu, Romania

Scientific Committee

Kiril Alexiev	Bulgarian Academy of Sciences, Bulgaria
Anca Andreica	Babeş-Bolyai University, Romania
Christopher Auer	University of Applied Sciences Landshut, Germany
Nebojsa Bacanin	Singidunum University, Serbia
Maryam Bagheri	Sheffield Hallam University, UK
Arndt Balzer	Technical University of Applied Sciences Würzburg-Schweinfurt, Germany
Costin Bădică	University of Craiova, Romania
Alina Bărbulescu	Ovidius University of Constanţa, Romania
Lasse Berntzen	University of South-Eastern Norway, Norway
Charul Bhatnagar	GLA University, India
Florian Boian	Babeş-Bolyai University, Romania
Peter Braun	Technical University of Applied Sciences Würzburg-Schweinfurt, Germany
Amelia Bucur	Lucian Blaga University of Sibiu, Romania
Steve Cassidy	Macquarie University, Australia
Camelia Chira	Babeş-Bolyai University, Romania
Nicolae Constantinescu	Lucian Blaga University of Sibiu, Romania
Ovidiu Cosma	Technical University of Cluj-Napoca, Romania
Cerasela Crişan	Vasile Alecsandri University of Bacău, Romania
Dan Cristea	Alexandru Ioan Cuza University, Romania
Gabriela Czibula	Babeş-Bolyai University, Romania
Daniela Dănciulescu	University of Craiova, Romania
Thierry Declerck	German Research Center for Artificial Intelligence, Germany
Lyubomyr Demkiv	Lviv National Polytechnic University, Ukraine
Gabriel Dimitriu	Grigore T. Popa University of Medicine and Pharmacy, Romania
Alexiei Dingli	University of Malta, Malta

Mika Tonder	Saimaa University of Applied Sciences, Finland
Claude Touzet	Aix-Marseille University, France
Milan Tuba	Singidunum University, Serbia & Sinergija University, Bosnia and Herzegovina
Dan Tufiş	Romanian Academy, Romania
Alexander D. Veit	Harvard Medical School, USA
Anca Vasilescu	Transilvania University of Braşov, Romania
Sofia Visa	College of Wooster, USA
Anca Vitcu	Alexandru Ioan Cuza University, Romania
Badri Vellambi	University of Cincinnati, USA
Xin-She Yang	Middlesex University London, UK

External Reviewers

Dana Vasiloaica	Atlantic Technological University, Ireland
Magda Gregorova	Technical University of Applied Sciences Würzburg-Schweinfurt, Germany
Radu Creţulescu	Lucian Blaga University of Sibiu, Romania
Daniela Marian	Technical University of Cluj-Napoca, Romania
Shyam B. Viswanathan	City University of New York, USA
Antoanela Luciana Naaji	Vasile Goldis Western University of Arad, Romania
Sabin Tabirca	University College Cork, Ireland & Transilvania University of Braşov, Romania
Filomena Ferrucci	University of Salerno, Italy
Mihai Gabroveanu	University of Craiova, Romania
Antoniu Pitic	Lucian Blaga University of Sibiu, Romania
Roxana Ardeleanu	Vasile Alecsandri University of Bacău, Romania
Maria Flori	Lucian Blaga University of Sibiu, Romania
Eva Tuba	Trinity University, USA & Singidunum University, Serbia
Vasile Berinde	Technical University of Cluj-Napoca, Romania
Gabriela Moise	Petroleum-Gas University of Ploieşti, Romania
Oana Adriana Ţicleanu	Lucian Blaga University of Sibiu, Romania
Mircea Muşan	Lucian Blaga University of Sibiu, Romania
Philipp Vaeth	Technical University of Applied Sciences Würzburg-Schweinfurt, Germany
Daniel Morariu	Lucian Blaga University of Sibiu, Romania

Sponsors

Academia Ardeleană Foundation

BIT Association

GSD

NTT Data

NXP

Pan Food

ROPARDO

Wenglor

Plenary Lectures

Ant Colony Optimization and Applications

Stefka Fidanova

Bulgarian Academy of Sciences, Bulgaria
stefka@parallel.bas.bg

Abstract. Collective intelligence allows animals to overcome difficulties and obstacles in nature, even when they do not have a high level of individual intelligence. Bees, ant colonies, birds, fish, etc. can be given as examples of group intelligence. This phenomenon has inspired scientists to imitate nature in their quest to solve complex problems coming from real life and industry. The unique behavior of ants in nature and their ability to always find the shortest path between their nest and a food source gives the idea for the creation of the ant method. Applications of ant colony optimization include the knapsack problem, the grid scheduling problem, the GPS surveying problem, the bioreactor modeling problem, wireless sensor network positioning, image edge detection, workforce planning, agricultural modelling and many others. The goal of this presentation is to show the diverse capabilities and applications of ant colony optimization.

Deep Generative Modelling - Beyond Language and Videos

Magda Gregorová

Technical University of Applied Science Würzburg-Schweinfurt, Germany
magda.gregorova@thws.de

Abstract. In this talk we will discuss deep generative models for other domains and purposes than generating seemingly realistic pieces of text, images or videos. We will explore a range of problems where we use deep unsupervised methods to learn high-dimensional data distributions for various downstream tasks. In particular, we will investigate the problem of modelling complex relational structures represented as graphs and the challenges these pose for generating new graph examples (e.g. for de novo drug design), we will see how generative models can be used for explaining decisions of independent predictor models, and we will look at some examples of industrial applications such as deep generative modelling for optimization of electrical engine designs. We will conclude with some more general thoughts about the current state of research in deep generative modelling.

Computing with Neuron-Like Devices: SNP Systems

Andrei Păun

University of Bucharest, Romania
apaun@fmi.unibuc.ro

Abstract. SNP systems are a type of distributed, parallel, neural-like computation model inspired by how biological neurons process information and communicate through spikes. Over the past decade, significant progress has been made in understanding neurons and the brain, partly due to two major initiatives: the Human Brain Project (Europe) and the BRAIN Initiative (US). We will discuss several findings related to SNP systems and their variants as computational models, many of which demonstrate Turing Universality, even under certain constraints. Recent research results and open questions will also be described, particularly concerning Spiking Neural P systems with communication on request (SNQ P systems). We have been able to design small, Turing universal SNQ P systems using a limited number of neurons. Specifically, a Turing universal generative SNQ P system (or acceptor and function computing) has been constructed with 12 neurons (11 for acceptor and 21 for function computing).

Intelligent Systems for Digital Images

Milan Tuba

Singidunum University, Serbia
Sinergija University, Bosnia and Herzegovina
tuba@ieee.org

Abstract. Many modern systems, such as self-driving cars, security apps, and automated diagnostics, rely heavily on digital image classification. Over the years, researchers have developed many techniques, but convolutional neural networks (CNNs) have transformed the field, greatly improving classification accuracy. Results of previously advanced research topics are now achievable with CNNS without a lot of effort or time. However, CNNs also introduced new challenges. One big issue is finding the best CNN architecture, which is really tough because there are so many variables to adjust, such as the number and type of layers, the number of neurons, kernel size, pooling type, optimization algorithm, padding, and stride. Each model should be fine-tuned for each problem. To handle this, an often-used method is a trial-and-error approach, or a method called grid search. Since this is an optimization problem, recent studies have also tried metaheuristics such as swarm intelligence algorithms. While these methods take a lot of time, they can significantly boost accuracy. This talk will explore the latest advancements and challenges in CNNs, such as tuning hyperparameters.

Contents

Intelligent Systems for Decision Support

How to Select Chief Digital Officer to Drive Digital Transformation: A Multiple Attributes Group Decision-Making Model

Daniela Borissova[1(✉)], Zornitsa Dimitrova[1], Naiden Naidenov[1], Magdalena Garvanova[2], Ivan Garvanov[2], and Radoslav Yoshinov[3]

[1] Institute of Information and Communication Technologies at the Bulgarian Academy of Sciences, 1113 Sofia, Bulgaria
{daniela.borissova,zornitsa.dimitrova, naiden.naidenov}@iict.bas.bg

[2] University of Library Studies and Information Technologies, 1784 Sofia, Bulgaria
{m.garvanova,i.garvanov}@unibit.bg

[3] Laboratory of Telematics at the Bulgarian Academy of Sciences, Sofia, Bulgaria
yoshinov@cc.bas.bg

Abstract. Increasingly, many companies are now creating Chief Digital Officer (CDO) positions to support their digital transformation. As a relatively new position, the responsibilities involved cover various aspects of the company's operations, the most significant of which is culture change. The CDO is expected to develop and implement a digital strategy to be integrated into a digital ecosystem. The CDO must be able to identify new technologies such as artificial intelligence, and the Internet of Things that are relevant to the company's operations and will lead to improving existing solutions. Considering the importance of this position, this paper proposes a CDO selection model. A distinctive feature of the model is the ability to integrate various objective and subjective characteristics of the candidates for this position. Thus, the proposed group decision-making model could aggregate experts' assessments according to their points of view about the importance of the characteristics of candidates. The numerical application of the proposed model is based on six technical skills and seven soft skills forming the general performance of the candidates, evaluated by a group of three experts. The results show that by aggregating properly different experts' opinions it is possible to determine the most suited candidate for the CDO position.

Keywords: Chief Digital Officer · Digital Transformation · Technical & Soft Skills · Group Decision-Making

1 Introduction

Organizations are in a continuous digital transformation that affects all spheres of business. Integration of digital technologies requires the development of digital culture and skills before investing in digital infrastructure and technology. This change has to be done

D. Simian and L. F. Stoica (Eds.): MDIS 2024, CCIS 2486, pp. 3–17, 2025.
https://doi.org/10.1007/978-3-031-87386-7_1

before reconfiguring business models [1]. This requires digital leaders to lead these initiatives, ensuring a smooth transition and that the organization remains competitive. Most digital transformation efforts are related to abandoning on-premises solutions and moving them to the cloud thus integrating all company data into one centralized system. To be successful every digital transformation must be started with a predetermined vision about the organization's goals – what problem will the digital transformation solve and how will digitalization improve the organization? Some empirical analyses show that transformation urgency and coordination needs predict Chief Digital Officer (CDO) presence [2]. The CDO position differs from supposedly related roles, such as the chief information officer, the chief data officer, the chief innovation officer, and the chief security officer [3–5]. Digital transformation must become a priority for senior management which has to define the concept of the corporate business strategy to be followed [6].

CDO leadership role demands building digital capabilities that contribute to digital acceleration, digital marketing, and digital harmonization related to digital innovation, data analytics, and customer engagement [7]. The responsibilities of CIOs are closely related to technical aspects and IT efficiency, while CDOs' responsibilities are oriented towards business and strategic aspects [8]. A comprehensive understanding of the emerging research on the phenomenon of CDO and its importance during the digital transformation process of organizations is presented in [9]. Another systematic large-scale study of the emergence, nature, and determinants of the CDO position is given in [10].

The recent research identified principal research tendencies of marketing related to digital marketing performance from the consumer point of view; the use of social networks as a marketing strategy; knowledge of consumer behavior; the impact of mass media campaigns through digital technologies, etc. [11]. The authors also mark principal tendencies in management like information management and performance; the design of flexible organizations capable of adapting to the business environment; the influence of Internet technologies and digitalization on logistics and supply management. Recent research shows that SMEs effectively leverage the resources and capabilities of artificial intelligence (AI) by mobilizing technology, coordinating production processes, and empowering skilled people [12]. Another study highlights that the key duties of a CDO are related to the ability to scan the external environment and look for unexpected digital trends [13]. Regardless of the level of digitization reached, every organization must be responsible in terms of cybersecurity by conducting periodic tests both of people and the relevant assets [14]. Since all these activities are among the CDO's responsibilities, the person for this position must be carefully selected.

The digital transformation relies on a skilled and high-performing IT team. The challenges facing organizations as a result of digital transformation are related to the lack of knowledge and skills in cyber security, understanding application architectures, integration of different software products, analysis, and data migration. This conclusion is drawn in a recent report that shows the hybrid nature of newly appointed CDOs, who possess business administration knowledge and technology skills, but also play a fundamental role in achieving strategic change [15]. That is why digital transformation experts are needed to organize such processes.

Digital transformation is difficult for many reasons, including complexity; organizational resistance; skills gap; legacy systems; missing strategy and vision; security and privacy concerns; resource constraints; evolving landscape; measuring success, etc. This is why this article aims to offer a transparent approach to evaluating the right person for a CDO to successfully organize a company's digital transformation. Due to the many tasks that CDO has to perform, the theory of multi-attribute utility can be used as a suitable tool, allowing for many factors to be taken into account. This approach refers to a predefined finite number of given alternatives that need to be evaluated to different weighted attributes to get a preference ranking [16]. On the other hand, considering the importance of this choice, it would be good to do it through group decision-making. Consequently, a group decision-making model is proposed for the selection of the CDO position that is capable of integrating both the technical and soft skills of applicants.

The rest of the article is organized as follows: Sect. 2 reviews the roles, responsibilities, and skills for the CDO position; Sect. 3 describes the proposed multiple-attribute model for group decision-making in the evaluation of the candidates for the CDO position; Sect. 4 contains numerical verification on the proposed model in the selection of CDO along with obtained results and discussion, and conclusions are drawn in Sect. 5.

2 Responsibilities, Tasks, Technical and Soft Skills for the CDO Position

Several key domains such as IoT, social media, mobile apps, artificial intelligence, augmented and virtual reality, the metaverse, and corporate digital responsibility are at the core of digital transformation [17]. All of these determine the complex nature of the job of the chief digital officer. Strong leadership abilities along with technical expertise can be considered the perfect mixture for the ideal candidate for the CDO position. The CDO is also known as a digital director or head of digital transformation that should cope with a variety of tasks summarized in 4 main directions as shown in Fig. 1.

Fig. 1. Responsibility of Chief Digital Officer.

Direction 1: Enabling digital transformation and value creation through new tools and technologies. This activity is related to cutting-edge technologies changing the way industries operate and related to AI, ML, IoT, blockchain, and cloud. Making predictions based on collected data from AI and ML is transforming the way businesses forecast thus enabling organizations to achieve higher efficiency [18]. Deep knowledge of these technologies can be considered as the basic prerequisite to successful CDOs. In-depth knowledge of these technologies can be considered a basic prerequisite for successful CDOs. To implement a digital transformation, knowledge in the field of business models, forecasting, and strategic planning, including a good knowledge of the organization's goals and finances, as well as prioritization of tasks is required [19]. All of these business skills would allow CDOs to make informed decisions about which technologies are best for the company. CDOs lean heavily on data to make decisions. Hence, this person must be able to handle a variety of data, aggregate data, and understand how to apply it to various digital initiatives.

Direction 2: Easing the company's transition to digital with customers and partners. Digital transformation is an evolution that is imperative for any company regardless of industry. Companies' online presence makes it possible for customers to communicate with businesses effortlessly and in a way that is preferred by customers. This is driven by customers demanding quick responses from businesses thus saving time and money, which the CDO has to deal with. Customers want to see the company's business in their favorite channels and mobile apps. Increased customer satisfaction and, accordingly, the realization of more economic profits can be achieved through digital transformation, as part of the activities focus on the user experience.

Direction 3: Communicating and persuading the importance of digitalization throughout the entire company. The big challenge of digital transformation is cultural change [20]. It covers all employees from management to new hires who need to be on the same wave. Everyone should be willing to make changes and learn new things daily.

Direction 4: Making digital a fundamental part of the organization's strategy. CDOs have to ensure the digital transformation meets business standards and achieves company goals. The ability to successfully lead a digital transformation depends to a large extent on the digital strategy defined by managers capable of creating a corporate culture. An effective strategy for managing this change is based on investigating and identifying the root causes of problems and building relationships with all stakeholders and employees [21].

To succeed in digital transformation, a company must develop digital change management skills and build a digital culture. The non-profit sector also faces the challenges of digital transformation [22]. It is shown that not-for-profit organizations could also implement digital transformation by following certain guidelines for strategic progress [23]. Therefore, the primary task of the CDO is to become a "digital expert" who helps the company to implement changes in the business so that it adapts to new technologies and consumer preferences, volunteer staff, clients, and donors. In this regard, the suitable person capable of realizing this process needs to comply with a variety of technical and soft skills.

Technical skills express specific abilities and skills that a person possesses and can demonstrate. For the CDO position, required technical skills include:

- Technology awareness. A CDO's core skills are in information technology as they utilize a variety of software tools and platforms to perform their role effectively. The CDO must be aware of the latest technologies and how they can be used to improve the company's operations. They must identify the most effective technology for specific business needs, identify the most effective ways to apply the technology, and explain how it can be used to increase productivity.
- Understanding of digital structures and products. The CDO should be able to test new technologies, to understand them, and to implement them in business. It should look for the next new digital trend and decide whether it's worth a company's attention.
- Degree in business, technology, or engineering. Having this type of qualification implies that the candidate will be familiar with tools, techniques, and frameworks that can enable and support digital innovation and strategy.
- Decision-making. CDOs rely on data to make well-informed decisions best suited to the company's goals. Complex decision-making challenges are related to the use of AI, digital transformation, and cyber security in the banking sector [24]. Therefore, CDOs are to be able to work with different data types, aggregate data, and propose proper decision-making models for various digital initiatives [25, 26].
- Project management. CDO uses digital information and advanced technologies to create business value. Digital transformation as a project requires CDO to be aware of the best practices and methodologies of Agile project management.
- Previous experience. The CDO is a relatively new leadership position and previous experience in a similar role could be an advantage.

Soft skills are related to communication, presentation skills, teamwork, innovative mindset, managing people, etc., and are not taught in school or college programs. Ideally, CDOs should have the perfect mix of technical expertise, business know-how, and leadership skills. A list of the main soft skills contributing to the competence of the CDOs is related to:

- Excellent communication skills. Both inside and outside the organization, the chief digital officer must be able to communicate digital processes and explain their benefits to everyone involved. Since the main task of CDO is to change the usual work processes by implementing new ways of doing things, open and clear communication is important.
- Leadership. The CDO is to be a leader capable of leading cross-functional teams in a dynamic and complex environment. The CDO directs and coordinates employees to achieve desired operational plans and results. He/she fosters a culture of continuous improvement and prioritizes the professional development of all members through his example of an inspirational leader.
- Time management skills. A CDO is a key person in an organization that undertakes complex digital transformation initiatives. In this regard, the position requires thorough planning, execution, and coordination over time. So, CDOs have to possess perfect skills in time management to guarantee on time the completion of projects, according to all deadlines and priorities, and last but not least to deliver results effectively.
- High sense of responsibility and perseverance. Implementing digital transformation involves complex and fundamental changes for the entire organization, which are

accompanied by many risks. As the CDO plays the role of an engine of change, he/she needs to be responsible and persistent enough to drive the transformation forward.

- Emotional intelligence. CDOs are to possess the ability to understand, use, and positively manage their own emotions. The successful CDO possesses an emotional intelligence that allows him to develop stable working relationships with employees, customers, and investors.
- Collaboration and teamwork. Collaboration skills are essential to work effectively with diverse stakeholders to achieve goals through communication, teamwork, and problem-solving. So, CDOs need to be able to handle different formal and informal activities to be able to coordinate and manage to work in different units and at different hierarchical levels [27].
- Strategic thinking. CDOs are to be able to build digitally driven business strategies and innovations. This skill helps skillfully monitor and strategically plan organizational efforts on long-term and impactful initiatives.

CDOs as senior C-level executives help the company create and execute digital initiatives that drive digital revenue, improve customer experience and employee engagement, and optimize operations while creating new business value.

The important positioning of the CDO due to his influence in the organization, motivates the need to select the right candidate, which can go beyond the scope of their previous and current successes. Thus, assessing potential in addition to current performance is critical to the successful recruitment of digital strategists and CDOs.

3 Group Decision Model in Selection of CDO

The basic formulation of multicriteria decision-making problem is usually expresses in the following matrix format [28]:

$$\begin{array}{c|c} & C_1 \;\ldots\ldots\ldots\; C_n \\ \hline \begin{array}{c} A_1 \\ \ldots \\ A_m \end{array} & \begin{bmatrix} r_{11}^{k} & \cdots & r_{1n}^{k1} \\ \vdots & \ddots & \vdots \\ r_{m1}^{k} & \cdots & r_{mn}^{k} \end{bmatrix} \begin{bmatrix} w_1^{k} \\ \ldots \\ w_n^{k} \end{bmatrix} \end{array} \tag{1}$$

where the alternatives (CDO candidates) are denoted by $A_1, \ldots A_m$, evaluation criteria are expressed by $C_1, \ldots C_n$, performance rating of A_m toward C_n from k-th expert is expressed by r_{mn}^{k} and w_n^{k} represents the weight the importance of n-th criterion given by k-th expert.

To make an effective selection of the most preferable candidate according to the entire group's opinions, the evaluations of each group member are to be considered with different importance. Using weighted coefficients, it is possible to consider different domain knowledge and experience of the expert with different importance in the final group decision. For this purpose, the following optimization model is proposed:

$$CDO^* = max\left\{\left(\alpha^1\begin{bmatrix} r_{11}^1 & \cdots & r_{1n}^1 \\ \vdots & \ddots & \vdots \\ r_{m1}^1 & \cdots & r_{mn}^1 \end{bmatrix}\begin{bmatrix} w_1^1 \\ \cdots \\ w_n^1 \end{bmatrix}\right) + \cdots + \left(\alpha^k\begin{bmatrix} r_{11}^k & \cdots & r_{1n}^{k1} \\ \vdots & \ddots & \vdots \\ r_{m1}^k & \cdots & r_{mn}^k \end{bmatrix}\begin{bmatrix} w_1^k \\ \cdots \\ w_n^k \end{bmatrix}\right)\right\}^i \tag{2}$$

subject to

$$\sum\nolimits_{k=1}^{K} \alpha^k = 1 \tag{3}$$

$$\sum\nolimits_{j=1}^{N} w_j = 1 \tag{4}$$

where α^k is weighted coefficient that expresses the level of expertise/importance of k-th expert. Given relations (3) and (4), the range of estimates must be between 0 and 1 where the bigger value means better performance of the evaluated criterion. This is because to obtain comparable values, all three parameters must have the same range of variation. The proposed model is capable of integrating other measurement approaches such as the Absolute Category Rating (ACR) method. This method could implement an overall quality rating scale with a five-point (Bad, Poor, Fair, Good, Excellent), or if necessary – a higher nine-point scale. No matter which of these scales will be used, it is imperative that the obtained results are normalized to the interval from 0 to 1. .. The tricky way is how to determine the coefficients in (3). One possible approach to cope with this is to use an integrated coefficient composed of the experience of experts in years and expertise of area competency to the particular problem as discussed in [29].

It should be noted that in the proposed model (2)–(4) all technical and soft skills are considered with equal importance. If it is needed to differentiate these skills additional coefficients about technical and soft skills are to be introduced.

To select the most appropriate candidate for the position of CDO we propose two types of evaluation criteria referring to the technical and soft skills. These two types of evaluation criteria are shown in Table 1.

To estimate the performance of the candidates listed above technical and soft skills in combination with the proposed group decision-making model will be used.

Table 1. Criteria for evaluation technical and soft skills of candidates for CDO.

#	Technical skills	#	Soft skills
t-1	Broad technological awareness	s-1	Communication
t-2	Understanding of digital structures and products	s-2	Leadership
t-3	Degree in business/technology/engineering	s-3	Responsibility and perseverance
t-4	Data analysis and problem-solving ability	s-4	Time management
t-5	Leadership of successful digital transformation projects	s-5	Emotional intelligence
t-6	Experience in managing and leading a digital team	s-6	Collaboration and teamwork
		s-7	Strategic thinking

4 Numerical Testing

Traditionally, the process of personnel hiring is based on several steps including screening and pre-selection, interviews, and assessment by answering some questionnaires, references, and background checks. To formalize these steps in a reasoned decision, a CDO selection case study of a small trading firm is considered, where a group of experts is selected to participate in the evaluation of the candidates. For this purpose, the evaluation criteria described above are used. Unlike soft skills, technical skills can be measured and therefore in this numerical testing, they are considered already assessed and available with normalized scores. For example, technical skills can be assessed by passing proper tests that will give information about technology awareness and understanding of digital structures and products. A degree in business, technology, engineering, or project management could be proved by corresponding certifications, etc.

For the numerical testing of the proposed model, a group of 3 experts was formed from a chief executive officer who formulated business objectives and made strategic decisions (E-1), human resources (E-2), and chief technology officer (E-3). For the position of CDO, 5 candidates have submitted documents and have to be ranked considering the aggregated score of the formed group of experts.

4.1 Input Data

The preliminarily determined evaluations of technical skills are done by using a scale from 0 to 100 where the bigger value means the better performance. Then the obtained scores were normalized in the range from 0 to 1 to be comparable with the other coefficients. These normalized evaluations concerning technical skills along with the corresponding weights for their importance according to experts' points of view are shown in Table 2.

Once the scores for technical skills and their importance coefficients are known, the next step is to assess soft skills. The evaluations of the soft skills from the conducted

Table 2. Evaluation score and weights for technical skills according to experts' points of view.

Technical skills	Candidates					Weights for technical skills importance		
	1	2	3	4	5	E-1	E-2	E-3
t-1	0.8	0.85	0.78	0.79	0.84	0.100	0.080	0.090
t-2	0.77	0.85	0.8	0.85	0.75	0.100	0.085	0.095
t-3	0.87	0.78	0.75	0.68	0.7	0.085	0.050	0.070
t-4	0.8	0.82	0.86	0.76	0.84	0.070	0.090	0.080
t-5	0.78	0.75	0.85	0.69	0.85	0.075	0.070	0.080
t-6	0.76	0.73	0.8	0.84	0.67	0.060	0.100	0.080

interview together with the coefficients for their importance according to each expert are shown in Table 3.

Table 3. Evaluation score and weights for soft skills of candidates from experts.

Soft skills	Candidates					Weights for soft skill importance
	1	2	3	4	5	
	E-1: Evaluations					
s-1	0.68	0.75	0.67	0.7	0.7	0.085
s-2	0.69	0.78	0.82	0.8	0.8	0.08
s-3	0.75	0.65	0.7	0.8	0.75	0.06
s-4	0.7	0.72	0.73	0.75	0.75	0.07
s-5	0.8	0.67	0.76	0.8	0.78	0.075
s-6	0.8	0.75	0.8	0.82	0.76	0.075
s-7	0.68	0.73	0.82	0.85	0.72	0.065
	E-2: Evaluations					
s-1	0.72	0.65	0.7	0.85	0.68	0.08
s-2	0.78	0.75	0.73	0.78	0.75	0.07
s-3	0.66	0.8	0.7	0.7	0.65	0.075
s-4	0.85	0.73	0.72	0.65	0.85	0.08
s-5	0.82	0.78	0.67	0.75	0.82	0.065
s-6	0.75	0.72	0.75	0.8	0.75	0.075
s-7	0.78	0.73	0.75	0.67	0.75	0.08

(*continued*)

Table 3. (*continued*)

Soft skills	Candidates					Weights for soft skill importance
	1	2	3	4	5	
	E-3: Evaluations					
s-1	0.72	0.74	0.82	0.72	0.78	0.07
s-2	0.68	0.72	0.75	0.67	0.82	0.09
s-3	0.72	0.75	0.7	0.75	0.85	0.065
s-4	0.7	0.72	0.8	0.76	0.72	0.07
s-5	0.8	0.67	0.75	0.85	0.77	0.06
s-6	0.75	0.75	0.75	0.82	0.75	0.07
s-7	0.75	0.68	0.65	0.78	0.82	0.08

Since the proposed model is based on group decision-making, 3 different scenarios are simulated. These scenarios represent 3 cases with different combinations of the importance of expert opinion, as shown in Table 4.

Table 4. Different cases for combining experts' opinion.

Cases	Weight for the importance of experts' opinion		
	E-1	E-2	E-3
Case-1	0.333	0.333	0.333
Case-2	0.20	0.35	0.45
Case-3	0.45	0.30	0.25

The first scenario expressed by Case-1 considers all experts within the group with equal importance, while the next two scenarios take into account experts' opinions with different importance. Case-2 expresses the situation in which expert E-3 is considered for the most qualified and therefore his opinion is considered the most important, followed by E-2 and E-1. In Case-3 the expert E-1 is the most important and its opinion is highly valuable, followed by E-2 and E-3. The input data from Tables 2, 3, and 4 are used to solve several tasks based on the proposed model (1)–(4). The results of the calculation using the formulated model and described input data are illustrated in Sect. 4.2.

4.2 Results Analysis and Discussion

The results of three cases of the relative importance of the experts' opinions regarding the combination of technical and soft skills are shown in Fig. 2.

Using Case-1 where the opinions of all experts are taken with equal importance, the best-suited candidate for the position of CDO is #3. He has the highest value of 0.7503

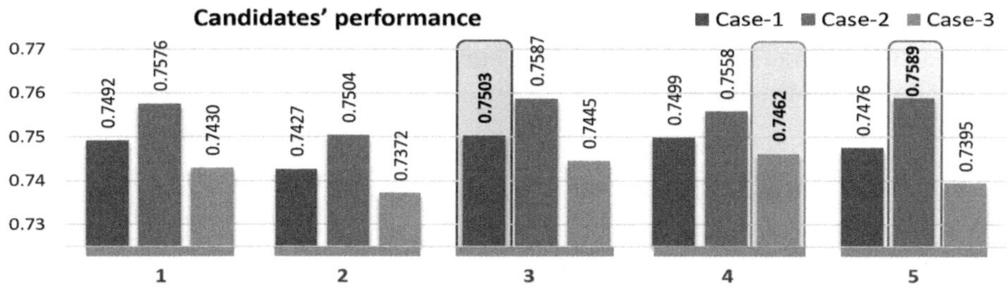

Fig. 2. Comparison of the simulated 3 different cases about the importance of experts' opinions.

compared to the others. When simulation the situation under Case-2 (see Table 4), with the predominate importance of E-3 the results show another suitable candidate under # 5 considering the aggregated preferences of all experts. In this case, the obtained highest value is 0.7589. Taking into account the situation determined via Case-3 from Table 4, the results show that the most appropriate candidate for the CDO position is #4 with an overall value of 0.7462. In a such way, it is possible to simulate different scenarios for the selection of the best-suited person considering different technical and soft skills evaluated by a group of experts.

It's interesting how each expert rated the candidates and who was their top choice according to their preferences. The answer to this question concerning the best candidate according to the individual preferences of the experts within the group is illustrated in Fig. 3.

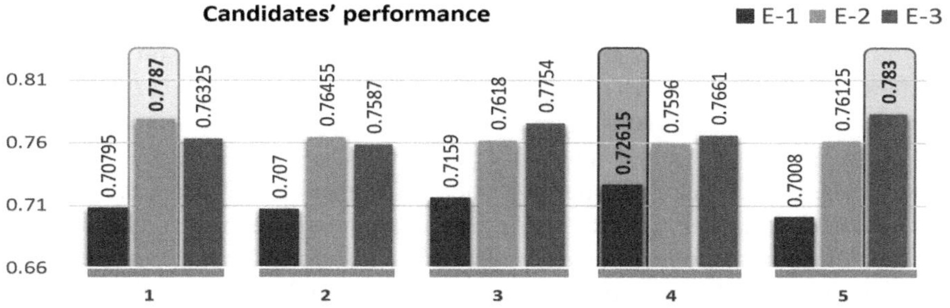

Fig. 3. Comparison between the individual selection for the position of CDO.

From Fig. 3 can be seen that expert E-1 selected the #4 (with an overall score of 0.72615) candidate as the best for the position of CDO. Expert E-2 determines #1 as the best candidate with a score of 0.7787. According to the point of view of the expert E-3, the best candidate is under # 5 with a score of 0.783. It could be seen, that different experts have different preferences about the best CDO candidate.

The use of weighted coefficients on the opinions of the experts can mitigate this imbalance in the preferences of each of the experts in forming the final group decision. These weighting coefficients can be determined using additional criteria such as previous experience, expertise in the specific field, etc. In the present experiment, they are treated as known in advance.

The comparison between the selected candidates in the group decision-making scenarios and the selected candidates according to the individual preferences of the experts in the group is shown in Fig. 4.

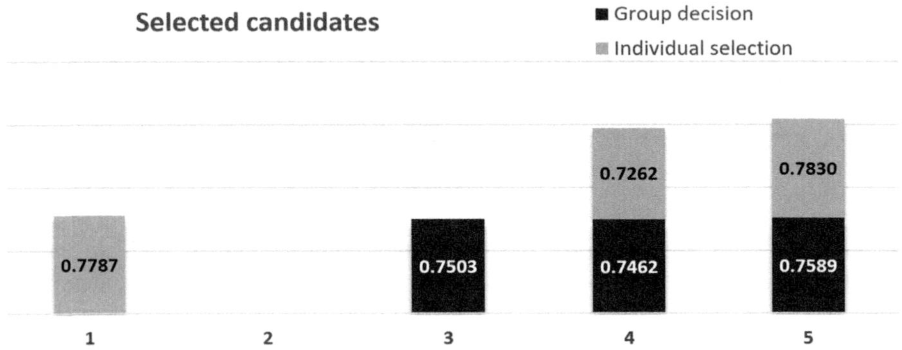

Fig. 4. Comparison between group decision making and individual choice for CDO candidates according to different experts.

It can be seen from Fig. 4 that there are 2 candidates (#4 and #5) that are the preferred choice of both individual experts and as a result of group decision making. Although candidate #1 was not among those selected in the group decision (see Fig. 2), he is the preferred choice according to expert E-2 point of view for the position of CDO. Therefore, by aggregating properly different experts' points of view it is possible to determine the most suited candidate for the CDO position.

Will the selected CDO candidate be successful in the selected position? An answer to this question can be obtained by analyzing the state of the company after the first/second year. Various indicators can be used for this purpose, such as the percentage of AI-enabled businesses; implementation of innovative solutions; new products or services; time to market for new offers, etc. Another point of view to measure success can be customer satisfaction, customer experience metrics, etc. This interesting research direction is a good continuation of the investigated topic, which is planned for future research.

The integration of such heterogeneous data can be applied also to other application areas where it is necessary to take into account information from different sources to form a common solution. For example, data from different sensors for multiple measurement variables to create comprehensive state information and data about the environmental state or current location, etc.

The proposed group decision-making model can be easily modified by adding or reducing some of the evaluation criteria as they are not strictly divided in the decision matrix (1). Moreover, it is possible to use a different number of experts to make the group decision. By changing some of the technical and/or soft skills it is possible to select suitable candidates for other or similar positions. In this way, it is possible to improve progress in digitization, which in turn contributes to better economic sustainability.

5 Conclusions

The article deals with the issues of digital transformation and more precisely with the selection of a leading person of digital transformation – the Chief Digital Officer. For this goal, the main responsibilities and tasks are described. Based on these responsibilities and tasks several technical and soft skills are identified as the most important. Taking into account the described problem specifics a multi-attribute group decision-making model to determine the most preferable candidate for the CDO position is formulated. The applicability of the proposed approach is numerically tested in the case of five candidates evaluated by a panel of three experts. Three different scenarios are simulated to investigate the influence of experts' opinions importance in the final group decision. The obtained results provide evidence regarding the applicability and practicality of the proposed model. Future research concerns the implementation of this approach in a web-based application along with research related to outcome indicators of CDO's achievement.

Acknowledgment. This study was funded by the project "Mathematical models, methods and algorithms for solving hard optimization problems to achieve high security in communications and better economic sustainability", KP-06-H52/7/19-11-2021 and by the project "Innovative Methods and Algorithms for Detection and Recognition of Moving Objects by Integration of Heterogeneous Data", KP-06-N 72/4/05.12.2023.

References

1. Brunetti, F., Matt, D.T., Bonfanti, A., De Longhi, A., Pedrini, G., Orzes, G.: Digital transformation challenges: strategies emerging from a multi-stakeholder approach. TQM J. **32**(4), 697–724 (2020). https://doi.org/10.1108/TQM-12-2019-0309
2. Firk, S., Hanelt, A., Oehmichen, J., Wolff, M.: Chief digital officers: an analysis of the presence of a centralized digital transformation role. J. Manage. Stud. **58**, 1800–1831 (2021). https://doi.org/10.1111/joms.12718
3. Dumeresque, D.: The chief digital officer: bringing a dynamic approach to digital business. Strateg. Dir. **30**(1), 1–3 (2014). https://doi.org/10.1108/SD-12-2013-0104
4. Singh, A., Klarner, P., Hess, T.: How do chief digital officers pursue digital transformation activities? The role of organization design parameters. Long Range Plan. **53**(3), 101890 (2020). https://doi.org/10.1016/j.lrp.2019.07.001
5. Borissova, D., Dimitrova, Z., Naidenov, N., Yoshinov, R.: Integrated approach to assessing the progress of digital transformation by using multiple objective and subjective indicators. In: Guizzardi, R., Ralyté, J., Franch, X. (eds.) Research Challenges in Information Science, RCIS 2022, LNBIP, vol. 446, pp. 626–634 (2022). https://doi.org/10.1007/978-3-031-05760-1_37
6. Saarikko, T., Westergren, U.H., Blomquist, T.: Digital transformation: five recommendations for the digitally conscious firm. Bus. Horiz. **63**(6), 825–839 (2020)
7. Tumbas, S., Berente, N., vom Brocke, J.: Three types of chief digital officers and the reasons organizations adopt the role. MIS Quart. Executive **16**(2), 121–134 (2017). https://aisel.aisnet.org/misqe/vol16/iss2/5
8. Ulrich, P., Lehmann, P.: Information is a key: Systematic literature review and empirical results on role configurations of chief information officers and chief digital officers. Corporate Bus. Strategy Rev. **4**(1), 87–98 (2023). https://doi.org/10.22495/cbsrv4i1art8

9. Christofi, M.: The role of chief digital officer: Critical insights into an emerging field and road map for future research. J. Bus. Res. **172**, 114390 (2024). https://doi.org/10.1016/j.jbusres.2023.114390
10. Kunisch, S., Menz, M., Langan, R.: Chief digital officers: an exploratory analysis of their emergence, nature, and determinants. Long Range Plan. **55**(2), 101999 (2022). https://doi.org/10.1016/j.lrp.2020.101999
11. Calderon-Monge, E., Ribeiro-Soriano, D.: The role of digitalization in business and management: a systematic literature review. RMS **18**, 449–491 (2024). https://doi.org/10.1007/s11846-023-00647-8
12. Peretz-Andersson, E., Tabares, S., Mikalef, P., Parida, V.: Artificial intelligence implementation in manufacturing SMEs: a resource orchestration approach. Int. J. Inf. Manage. **77**, 102781 (2024). https://doi.org/10.1016/j.ijinfomgt.2024.102781
13. Berbel-Vera, J., Palanca, M.B., Gonzalez-Sanchez, M.B.: Key CDO functions for successful digital transformation: insights from a Delphi study. Technol. Forecast. Soc. Chang. **181**, 121773 (2022). https://doi.org/10.1016/j.techfore.2022.121773
14. Blagoev, I.: Cyber security threats in the public hosting services. In: Tagarev, T., Stoianov, N. (eds.) Digital Transformation, Cyber Security and Resilience. DIGILIENCE 2020. Communications in Computer and Information Science, vol. 1790, pp. 63–72 (2024). https://doi.org/10.1007/978-3-031-44440-1_10
15. Culasso, F., Gavurova, B., Crocco, E., Giacosa, E.: Empirical identification of the chief digital officer role: a latent Dirichlet allocation approach. J. Bus. Res. **154**, 113301 (2023). https://doi.org/10.1016/j.jbusres.2022.113301
16. Jansen, S.J.T.: The multi-attribute utility method. In: Jansen, S., Coolen, H., Goetgeluk, R. (eds.) The Measurement and Analysis of Housing Preference and Choice. pp. 101–125 (2011). https://doi.org/10.1007/978-90-481-8894-9_5
17. Paul, J., et al.: Digital transformation: a multidisciplinary perspective and future research agenda. Int. J. Consum. Stud. **48**(2), (2024). https://doi.org/10.1111/ijcs.13015
18. Koliousis, I., Al-Surmi, A., Bashiri, M.: Artificial intelligence and policy making; can small municipalities enable digital transformation? Int. J. Prod. Econ. **274**, 109324 (2024). https://doi.org/10.1016/j.ijpe.2024.109324
19. Rego, B.S., Jayantilal, S., Ferreira, J.J., Carayannis, E.G.: Digital transformation and strategic management: a systematic review of the literature. J. Knowl. Econ. **13**, 3195–3222 (2022). https://doi.org/10.1007/s13132-021-00853-3
20. Burchardt, C., Maisch, B.: Digitalization needs a cultural change – examples of applying agility and open innovation to drive the digital transformation. Procedia CIRP **84**, 112–117 (2019). https://doi.org/10.1016/j.procir.2019.05.009
21. Mehta, N., Mehta, A., Hassan, Y., Buttner, H., RoyChowdhury, S.: Choices in CDO appointment and firm performance: moving towards a stakeholder-based approach. J. Bus. Res. **134**, 233–251 (2021). https://doi.org/10.1016/j.jbusres.2021.05.016
22. Nahrkhalaji, S. S., Shafiee, S., Shafiee, M., and Hvam, L.: Challenges of digital transformation: the case of the non-profit sector. In: IEEE International Conference on Industrial Engineering and Engineering Management (IEEM), pp. 1245–1249 (2018). https://doi.org/10.1109/IEEM.2018.8607762
23. Brink, H., Packmohr, S., Vogelsang, K.: Fields of action to advance the digital transformation of NPOs – Development of a framework. In: Buchmann, R.A., Polini, A., Johansson, B., Karagiannis, D. (eds.) Perspectives in Business Informatics Research. BIR 2020. Lecture Notes in Business Information Processing, vol. 398, pp. 82–97 (2020). https://doi.org/10.1007/978-3-030-61140-8_6
24. Rodrigues, A.R.D., Ferreira, F.A.F., Teixeira, F.J.C.S.N., Zopounidis, C.: Artificial intelligence, digital transformation and cybersecurity in the banking sector: a multi-stakeholder

cognition-driven framework. Res. Int. Bus. Finan. **60**, 101616 (2022). https://doi.org/10.1016/j.ribaf.2022.101616
25. Borissova, D., Dimitrova, Z., Dimitrov, V., Yoshinov, R., Naidenov, N.: Digital transformation and the role of the CIO in decision making: a comparison of two modelling approaches. In: Saeed, K., Dvorský, J. (eds.) Computer Information Systems and Industrial Management, CISIM 2022, LNCS, vol. 13293, pp. 93–106 (2022). https://doi.org/10.1007/978-3-031-10539-5_7
26. Stoyanova, K., Guliashki, V.: Group drop of sustainability: trade-off solutions between low returns and portfolio stability. Comput. Inf. **4**(1), 13–19 (2024)
27. Singh, A., Hess, T.: Chapter 9 - how chief digital officers promote the digital transformation of their companies. In: Galliers, R.D., Leidner, D.E., Simeonova B. (eds.) Strategic Information Management (2020). https://doi.org/10.4324/9780429286797
28. Shih, H.-S., Shyur, H.-J., Lee, E.S.: An extension of TOPSIS for group decision making. Math. Comput. Model. **45**(7–8), 801–813 (2007). https://doi.org/10.1016/j.mcm.2006.03.023
29. Borissova, D.: A group decision making model considering experts competency: an application in personnel selections. Comptes rendus de l'Academie Bulgare des Sciences **71**(11), 1520–1527 (2018)

Queues Detection and Skiers Distribution in the Ski Resorts

Adam Gonšenica[2], Andrej Mihálik[1](✉), and Roman Ďurikovič[1]

[1] Faculty of Mathematics, Physics and Informatics, Comenius University, Bratislava, Slovakia
andrej.mihalik@fmph.uniba.sk

[2] Faculty of Management, Comenius University, Bratislava, Slovakia

Abstract. This paper investigates the synergy between Multi-Agent-based Simulation (MAS) and AI-powered queue estimation to understand the complex dynamics of skier behavior and queue formation at ski lifts. We propose a novel approach where groups of skiers are modeled as autonomous agents within the MAS framework. These agents make decisions based on factors such as preferred slopes and runs, and tolerance for waiting in lines. The simulation environment meticulously replicates the ski area layout, encompassing slopes, lifts, and other relevant features. Factors such as lift capacity and travel time are also incorporated to ensure realistic modeling.

To enhance the accuracy of queue predictions, real-time data is obtained through an AI model trained on labeled images from security cameras around the ski lifts. The real-time queue length estimates derived from the AI model are then seamlessly integrated into the MAS simulation. This integration empowers the MAS with a more dynamic and data-driven understanding of queue formation throughout the day, leading to more accurate predictions of congestion patterns and wait times.

By leveraging the insights gleaned from the MAS simulations, operators can make more informed decisions regarding lift operations and strategic layout modifications. Optimized lift usage based on these predictions can potentially lead to reduced waiting times for skiers, significantly enhancing their overall experience. In conclusion, this paper presents a powerful framework that combines MAS and AI-powered queue estimation to offer ski resorts a valuable tool for optimizing skier experience and operational efficiency.

Keywords: Multi-agent simulations · Queues · Skier behavior

1 Introduction

Understanding and modeling skier behavior within ski resort is essential to optimize safety, enhance visitor experience, and improve overall resort management.

By simulating skier movements, managers can identify high-risk areas, optimize resource allocation, and tailor services to meet visitor needs. This paper

D. Simian and L. F. Stoica (Eds.): MDIS 2024, CCIS 2486, pp. 18–32, 2025.
https://doi.org/10.1007/978-3-031-87386-7_2

explores the significance of employing autonomous agents for skier behavior modeling, emphasizing its role in efficiency improvement and customer satisfaction within ski resort environments.

Previous research has shed light on the complexities of skier behavior and queue formation within ski resorts, aiming to optimize guest experience and resort operations. Barras et al. [1] introduced the Ski-Optim model, a multi-agent simulation approach that leverages RFID data from ski tickets to analyze skier movement within lift networks and ski runs. Validated in Verbier, Switzerland, this model accurately simulates queue formation and provides valuable insights for crowd management [1]. Furthermore, Delibašić et al. [2] explored skier behavior using sensor data and matrix factorization. Their work highlights the underutilized potential of RFID data in understanding skier movement patterns and preferences. While Delibašić et al. [2] successfully identified patterns in skier transportation, they acknowledge limitations due to incomplete data. Addressing these limitations and expanding such studies across multiple resorts can provide richer insights for optimizing skier experience.

Poulhes and Mirial [5,6] introduced Dynaski, a novel multi-agent simulation tool designed from a management and planning perspective. Dynaski allows for simulating and comparing different management scenarios to assess their impact on skier comfort and waiting times. This approach integrates various data sources for calibration, addressing the limitations of existing research in this field. They collected original data through field surveys (understanding skier mobility and group dynamics) and GPS data (estimating speed distributions) to complement traditional lift swipe data used for model input and calibration. A successful demonstration in La Plagne, one of France's largest ski areas, showcased Dynaski's potential for real-world use. The simulation revealed that constructing new housing near the area could lead to an influx of 1620 skiers and a 12% increase in average waiting times. While acknowledging the need for further validation with more extensive data and implemented projects, the authors highlight Dynaski's applicability not only to various ski resorts, but also to other leisure places and theme parks.

Pullman and Thompson [7] explored various strategies for optimizing queue times and customer experience at a ski resort through simulation modeling. Their findings revealed that simply increasing lift capacity without informing skiers about queue times can be counterproductive. Distributing demand across days or expanding ski trails had minimal impact. Conversely, providing real-time queue information at the resort emerged as the most effective and cost-efficient method, significantly reducing wait times and potentially improving customer satisfaction.

Additionally, Delibašić et al. [3] conducted a comprehensive review of research on ski lift transportation data. They highlight the underutilized potential of this data for ski resort management. Their review identifies three key research directions: 1) clustering skier lift data to understand skier behavior patterns, 2) exploiting ski lift data for ski injury research and prevention, and 3) congestion analysis in ski resorts. However, Delibašić et al. also emphasize that ski resorts have not yet opened up their data, hindering further advancements. This untapped potential, if harnessed, could significantly improve decision-making processes, operational efficiency, and guest satisfaction.

Long wait times at ski lifts can significantly deteriorate the guest experience at ski resorts. Optimizing lift operations and skier distribution requires a deeper understanding of complex skier behavior and queue formation dynamics. This paper proposes a novel framework that integrates Multi-Agent Simulation (MAS) with AI-powered queue estimation. The following sections will delve into the methodology and implementation details of our approach. We present results from empirical validation and case studies that demonstrate the effectiveness of this framework in optimizing skier flow and reducing queue times. By empowering ski resorts with real-time data and AI-driven insights, this research has the potential to revolutionize ski resort management practices and significantly enhance the overall skiing experience.

2 Image Processing for Queue Length Estimation

To bridge the gap between simulated and real-time queue dynamics, especially in scenarios where zoomed-out camera angles hinder individual skier detection, this section explores an alternative image processing approach. However, the success of this approach depends on a crucial preliminary step of data annotation.

Prior to utilizing a Convolutional Neural Network (CNN) based model for object detection, a comprehensive dataset of images annotated for the specific task is essential. In our case, this dataset consists of images captured from the ski resort's cameras, encompassing various queue lengths and configurations. Each image undergoes meticulous labeling by human annotators to identify the ground truth location of the queue within the frame. This annotation process typically involves drawing bounding boxes around the queue regions.

This annotated data serves as the foundation for training the CNN model. By analyzing these labeled images, the model learns to recognize the distinctive features that differentiate queues from other elements in the image. Through this iterative learning process, the CNN model refines its ability to accurately detect queues in unseen images.

With a prepared data set, we leverage a Faster R-CNN-RESNET50-FPN model, readily available within the PyTorch library [4]. This well-regarded model incorporates a ResNet-50 backbone for feature extraction and a Feature Pyramid Network (FPN) to address the challenge of detecting objects at various scales [8]. In our application, the model is used to identify the total mass of skiers forming the queue, rather than individual skiers (see Fig. 1).

To train the CNN model, we employed a dataset of annotated images encompassing various queue configurations. The model was trained for 10 epochs. During this training, the loss function decreased significantly, dropping from 0.09 to 0.043 (see Fig. 2). This substantial decrease indicates the model's improving ability to distinguish queue regions within the images. We opted to stop training after 10 epochs based on the shape of the loss curve, as it suggested further

Fig. 1. Illustration of queue detection using bounding box.

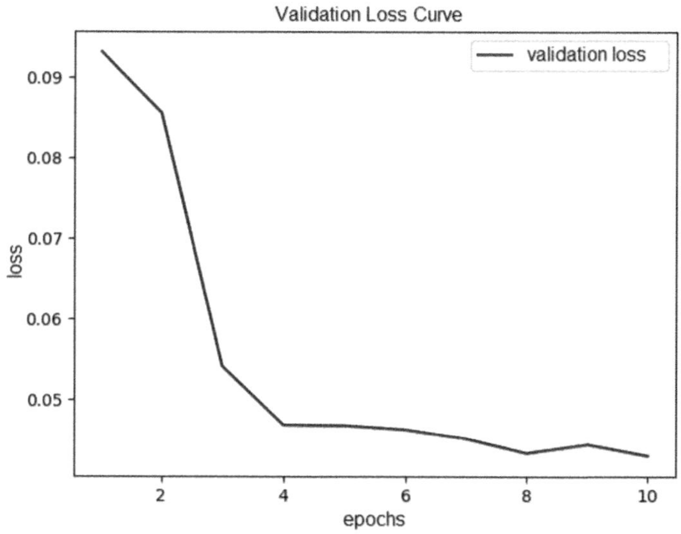

Fig. 2. Validation loss curve for the queue detection model. The loss function significantly decreases from 0.09 to 0.043 over 10 epochs.

training would not yield significant improvements to the model's performance on unseen camera images.

Based on the queue detection, the model generates a bounding box around the identified queue region. The width (w_b) and height (h_b) of this bounding box, measured in pixels, serve as inputs for estimating the queue length. To convert these pixel values into a waiting time attribute (T_w in minutes) usable within the MAS simulation, the following function is applied:

$$T_w = \begin{cases} 0, & \text{if } w_b < 100 \\ \frac{w_b}{100}, & \text{if } w_b > 100 \wedge h_b < 70 \\ \frac{w_b}{100} \cdot \left(1 + \frac{h_b - 70}{10}\right), & \text{otherwise.} \end{cases} \quad (1)$$

When the bounding-box width is less than 100 pixels, the queue length attribute is set to 0, indicating a short queue. For wider queues, the bounding box width is divided by a factor of 100 (empirically determined for the specific cameras deployed at the ski resort) to obtain a base queue length in minutes. This base value is then adjusted by a factor that considers the height of the bounding box. If the height exceeds 70 pixels, this suggests that the queue extends beyond the designated corridor. The adjustment factor increases the estimated waiting time based on the severity of this overflow (i.e., how much larger h_b is compared to 70). This real-time queue length information is then seamlessly integrated into the MAS simulation, enabling it to adapt to the evolving congestion patterns throughout the day and enhance the accuracy of its queue predictions.

The conversion factors used to estimate queue length from bounding box dimensions are empirically derived for the specific camera setup, which limits their generalizability. Incorporating known standard objects within the scene, such as fence heights, could potentially automate calibration, but accurate real-time data collection under varying environmental conditions remains challenging. To enhance model robustness, ground truth data with manually labeled queue lengths and corresponding bounding-box dimensions is essential for continuous calibration and refinement.

To validate the accuracy of the queue length estimation, we compared real-world measurements of queue lengths with the corresponding estimates generated by the model based on camera images. This comparison encompassed various queue configurations and time periods throughout the day. The results are presented in a bar graph (Fig. 3), which visually compares the model's estimated queue lengths with actual measurements.

3 Multi-agent-Based Simulation

Multi-Agent-based Simulation (MAS) is a computational modeling approach that simulates complex systems by representing individual agents and their interactions within a given environment. Each agent operates autonomously, making decisions based on predefined rules or objectives. MAS is used across various fields to study emergent phenomena, understand system dynamics, and analyze decentralized systems' behaviors and collective outcomes.

We use the Mesa-Geo [9] extension (version 0.4.0) for the Mesa agent-based modeling framework. This enables us to model the ski resort environment, including the terrain and the behavior of skiers, for analyzing skier behavior.

To improve efficiency, we can explore approximating large groups of skiers with similar characteristics and behaviors as a single agent. This technique, known as agent aggregation, can be particularly useful for representing groups with a common goal, such as a ski lesson group following an instructor. To

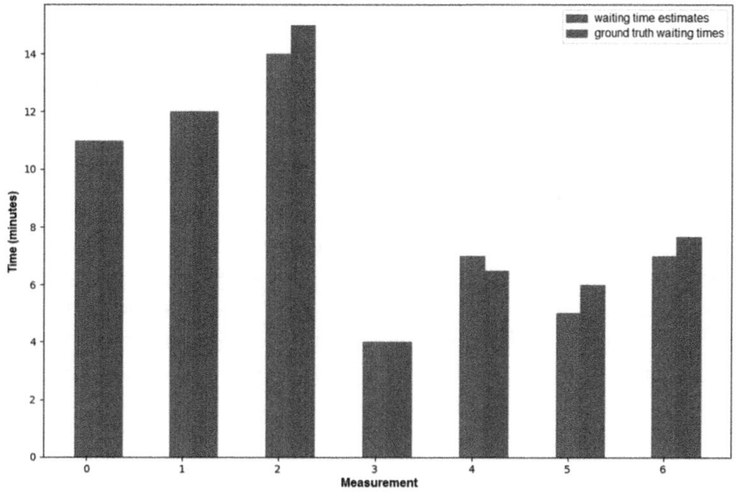

Fig. 3. This bar graph compares ground truth waiting times in minutes, measured on-site during peak hours, with the corresponding waiting time estimates generated by the model based on camera images.

balance complexity and efficiency, we model individual skiers with a moderate level of detail, capturing their skill level, preferred terrain, and basic decision-making for route selection. This allows us to analyze broad movement patterns and congestion at lifts while maintaining simulation performance. It's important to note that this approach might not capture the full spectrum of individual skier behavior, such as advanced skiing techniques.

We acquired a high-resolution Digital Elevation Model (DEM) of the ski resort terrain in GeoJSON format. These data provide the base map for our simulation environment, including attributes for each slope segment, such as open/closed status and difficulty level (e.g., beginner, intermediate, advanced). This information allows for a more realistic simulation of skier behavior as they navigate the resort based on ability and slope availability. The DEM data itself incorporate lift latitudes for elevation changes. However, information on the travel time to each lift was not available. To address this, we manually measured the lift travel times to accurately represent travel duration within the simulation.

We retrieved historical data from the Skidata-Cosima data warehouse, focusing on information relevant to skier behavior. This data includes the total number of tickets sold, the distribution of skiers between lifts based on turnstile data, and lift capacities per hour. This information helps us calibrate the simulation by matching simulated skier volume with real-world data. Skidata-Cosima data were preprocessed to remove outliers and ensure consistency with the chosen simulation time frame. The DEM and preprocessed Skidata-Cosima data are then integrated with the Mesa-Geo environment. The DEM defines the geographical space, while the skier data inform the number and distribution of simulated agents within the resort.

Modeling skier movement within a ski resort requires representing the complex network of interconnected runs and lifts. While a single directed acyclic graph might be ideal for one-way paths, it is not suitable in this scenario due to the presence of lifts, which create cycles within the network. To address this challenge, we can leverage the division of the ski resort into smaller, manageable sections based on factors such as difficulty level or geographic zones. Each section can then be represented by an oriented graph. The nodes within these graphs represent the start and end points of runs or slopes, along with any key intersections. Smaller directed edges connect nodes representing the downhill direction for skiers on runs or slopes. These edges can have attributes such as difficulty and length. Long segments represent the path of lifts (see Fig. 4). This approach offers a clear and manageable way to represent each section while enabling efficient pathfinding and skier movement simulation within Mesa-Geo.

Fig. 4. Oriented graph representing the ski resort network: Blue circles represent the vertices of the polyline, which defines the path that the autonomous agents navigate. Arrows indicate the direction of movement for the agents. (Color figure online)

Although a more detailed representation of the network could incorporate every possible connection point within the designated rectangular area (see

Fig. 5), such an approach would not realistically reflect skier movement on ski runs. Skiers mainly follow designated downhill paths with specific gradients. To balance accuracy with model complexity, we opted for a simplified approach that focuses on connecting key points within the small area, acknowledging that this introduces a slight deviation from the true network structure. This simplification allows for a more manageable model while still capturing the essential aspects of skier movement and network connectivity.

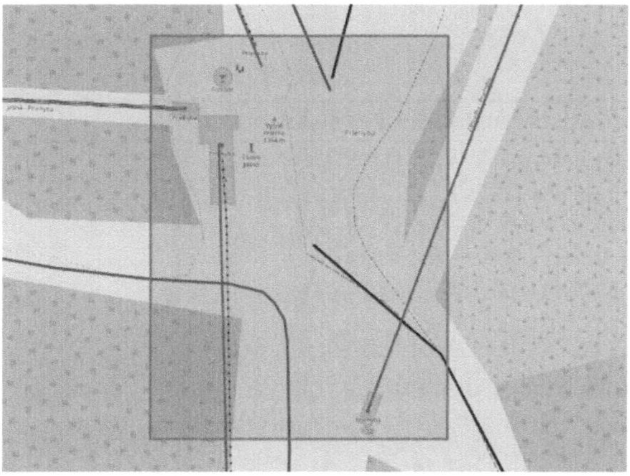

Fig. 5. Example of the rectangular area in the map, where we connect endpoints of the lift and slopes into one vertex.

3.1 Reparametrization for Travel Time Accuracy

Although our base model utilizes an oriented graph with segments representing runs or slopes, further refinement can be explored to enhance travel-time accuracy. This involves reparametrizing the segments based on average speeds associated with different difficulty levels or run types. This approach requires obtaining reliable data on average speeds and introduces additional complexity to the agent movement logic within the Mesa-Geo environment.

Our agent movement model utilizes the *RedistributeVertices* function (see Algorithm 1) to enhance travel-time accuracy within the Mesa-Geo environment. This function takes two inputs: a list G representing the original path vertices and a list S containing the average skiing speed for each segment of the path. The function iterates through the path segment-by-segment. For each segment, it compares the remaining distance with the product of the segment's average speed S_i, and the time interval of a single simulation step. If the remaining distance is less, the entire segment is covered in that step. Otherwise, the covered distance is determined based on S_i and the remaining rest time, r, from the previous step. The rest time is then reset, and a new vertex is inserted into an output list, P, at a location along the path calculated proportionally based on the covered distance, c, and the original path length. This location is obtained using the *Interpolate* function, which determines the vertex position on the polyline based on a parameter between 0 and 1. In our Python implementation, we leverage the LineString data structure and its built-in interpolation method for this calculation. This process continues until all segments are processed. Finally, the last point of the original path is added, and a new polyline is created from the list of vertices in P. This reparameterized path considers the skiers' average speeds on different segments, leading to more accurate travel times within the simulation.

Algorithm 1. Redistribute Vertices with Speeds

Require: Polyline geometry G, List of speeds S
Ensure: Polyline geometry P
  function RedistributeVertices(G, S) : P
  Initialize $c \leftarrow 0$, $r \leftarrow 0$, $i \leftarrow 0$, $k \leftarrow 0$
  Initialize a list $P \leftarrow \{G_0\}$
  while $i < |G| - 1$ **do**
    $\boldsymbol{l} \leftarrow G_{i+1} - G_i$
    if $|\boldsymbol{l}| - k < S_i \times (1 - r)$ **then**
      $c \leftarrow c + |\boldsymbol{l}| - k$
      $r \leftarrow \frac{(|\boldsymbol{l}| - k + S_i \times r)}{S_i}$
      $i \leftarrow i + 1$
      $k \leftarrow 0$
    else
      $c \leftarrow c + S_i \times (1 - r)$
      $k \leftarrow k + S_i \times (1 - r)$
      $r \leftarrow 0$
      $P \leftarrow P \cdot \{Interpolate(G, \frac{c}{|G|})\}$ {Concatenate with new vertex}
    end if
  end while
  $P \leftarrow P \cdot \{G_{|G|-1}\}$ {Concatenate with new vertex}
  return P
  end function

3.2 Agent Design and Route Choice

While the ideal scenario might involve modeling each skier as a distinct agent, computational limitations or the desire to focus on group dynamics may necessitate alternative approaches. In such cases, we can model a small group of skiers with similar characteristics, such as skill level and preferred terrain, as a single agent within the MAS framework. This agent's decision-making process would then incorporate a weighted average of individual preferences within the group it represents. This approach allows for a balance between model complexity and capturing essential group dynamics, particularly when simulating the behavior of small, cohesive groups of skiers.

For a simpler approach, agent route choice can leverage a popularity-based heuristic. This method focuses on selecting routes that are generally favored by skiers with the resort. Real-time data on queue lengths, obtained through an AI model, are used to estimate waiting times at different slopes. The queue is modeled as a first-in, first-out (FIFO) stack within the MAS. While it does not incorporate individual preferences within the group, it offers a computationally efficient way to guide group agents towards generally favored routes.

The flowchart in Fig. 6 shows the decision-making process of a simulated group agent within a ski resort environment. The agent starts at a "home" state and determines if it's time to enter the resort. Based on this decision, the agent might choose a lift. The system then computes the route to the chosen lift or the end of the planned skiing route. Throughout the simulation, the agent encounters decision points, such as the presence of a nearby lift, which influence its path and behavior.

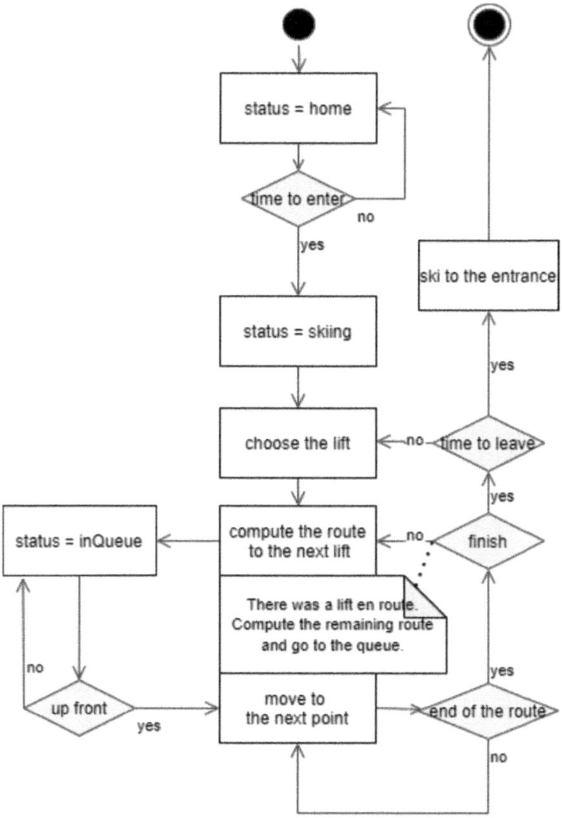

Fig. 6. State diagram for the simulated ski agent's decision making process.

3.3 Lift Popularity Estimation with Kernel Density Estimation for Agent Route Choice in Ski Resorts

We investigate the utilization of Kernel Density Estimation (KDE) with a Gaussian kernel within a Multi-Agent Simulation (MAS) framework to estimate the probability of skiers entering a turnstile at a particular time. This estimated probability, derived from historical turnstile data, impacts various factors such as the timing of skiers' entrance into the ski resort.

We analyze historical data on turnstile passes collected at each lift over a single day. These data represent a series of observations for particular turnstile l, denoted as $X_{l,1}, X_{l,2}, ..., X_{l,N}$, where N represents the total number of turnstile passes during the day. Each observation, $X_{l,i}$, signifies the time (in minutes) a skier used the turnstile l.

Our objective is to estimate the probability density function (PDF) denoted as $f_l(x)$, which describes the likelihood of encountering a specific turnstile l at a particular time x. To achieve this, we employ a Gaussian kernel function $k(x)$, with a bandwidth $h = 0.5$ and a center at zero mean ($\mu = 0$). This means that the Gaussian distribution is centered at zero, serving as a reference point for the kernel function's shape.

The Gaussian kernel function is defined as:

$$k(x) = \frac{1}{\sqrt{2\pi h^2}} \exp\left(-\frac{x^2}{2h^2}\right) \tag{2}$$

The estimated PDF using a Gaussian kernel can be expressed as:

$$f_l(x) = \frac{1}{Nh} \sum_{i=1}^{N} k\left(\frac{x - X_{l,i}}{h}\right) \tag{3}$$

Here, the kernel function is centered at each historical pass time $X_{l,i}$ and the sum is normalized by the total number of observations N. Figure 7 visually depicts the estimated PDF (smooth curve) alongside a histogram (bar chart) of the historical turnstile data.

For practical applications, we can approximate the normalization constant, K_t, for each time window:

$$K_t \approx (N \cdot h)^{-1/d} \tag{4}$$

where d is the dimensionality (here, $d = 1$ as we are dealing with a single variable). This approximation allows us to directly estimate the probability $P(l|t)$ of encountering skiers at lift entrance l during the time window t using the estimated density:

$$P(l|t) = K_t \cdot f_l(t) \tag{5}$$

To prioritize efficiency, the estimated probability of encountering any number of skiers at each lift along a route is integrated into the decision-making process. In this case, the lift with the highest overall probability is chosen. It's important

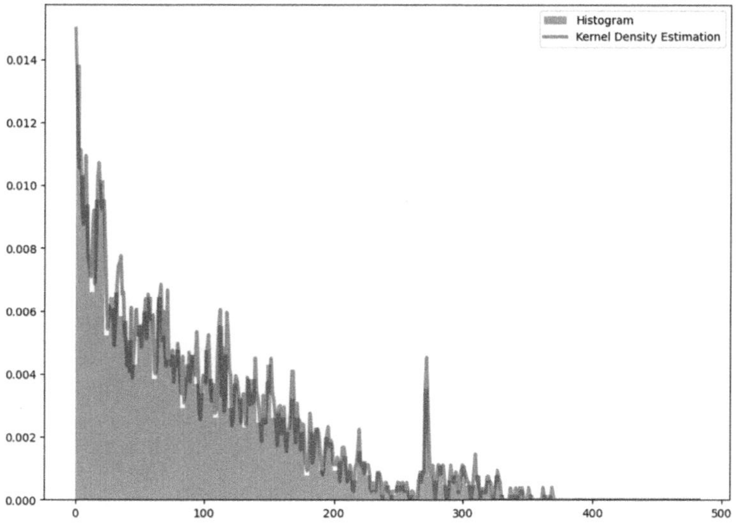

Fig. 7. Visualization of Kernel Density Estimation. The smooth curve represents the estimated probability density function. The blue bars represent the histogram of historical turnstile data collected over the same time window. The x-axis represents time in minutes (minutes past the start of the observation period), and the y-axis represents the frequency (or probability density) of skiers. (Color figure online)

to note that this method doesn't directly consider the specific number of skiers, but rather prioritizes lifts with a generally higher usage rate.

4 Results

To validate the simulation's effectiveness, real-world data was used. Three days with good weather and high skier volume were chosen from February 2024. The turnstile data from these days provided the reference for skier numbers entering the simulated environment. The MAS simulation was run ten times, with each run incorporating the same number of skiers as recorded by the turnstiles on the corresponding validation day. The results presented in this section represent the average outcome of these ten simulation runs.

Figure 8 illustrates a comparison of skier arrivals at the resort throughout the day, presented in hourly intervals. The real data is depicted in the dashed line, while the simulation results are shown in the solid line. The graph reveals a good fit between the real and simulated arrival patterns. Notably, both the morning influx and the distribution of arrivals throughout the rest of the day are captured with reasonable accuracy by the simulation. This successful replication of arrival patterns strengthens the validity of the MAS in representing real-world guest behavior.

Figure 9 delves deeper by comparing skier arrivals at a specific turnstile, located near the junction of a lift and a popular slope. Similar to the overall

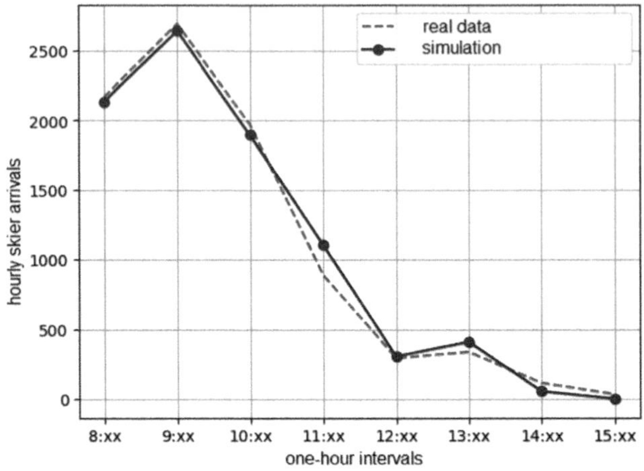

Fig. 8. Hourly skier arrivals at the resort. Real data is depicted by a dashed line, while simulation results are represented by a solid line.

arrival trend (Fig. 8), the graph depicts real data in the dashed line and simulation results in the solid line. The morning influx and peak hours are captured with good accuracy. However, the simulated afternoon decline is less pronounced, likely reflecting the possibility of skiers taking breaks for refreshments at the resort's restaurants and buffets, which might not be perfectly captured by the simulation. This highlights an area for potential improvement in future iterations of the MAS.

To gain further insights into skier behavior, we calculated the average number of times each skier passed through the turnstiles. This metric was obtained by dividing the total number of turnstile activations by the total number of skiers. Analyzing this data can reveal aspects like the frequency of skiers taking multiple runs on different slopes throughout the day. The results are presented in Table 1.

Table 1. Average turnstile passes per skier (Real vs. Simulated Data).

Simulated Data	8.99 turnstile passes/skier
Real Data	9.01 turnstile passes/skier

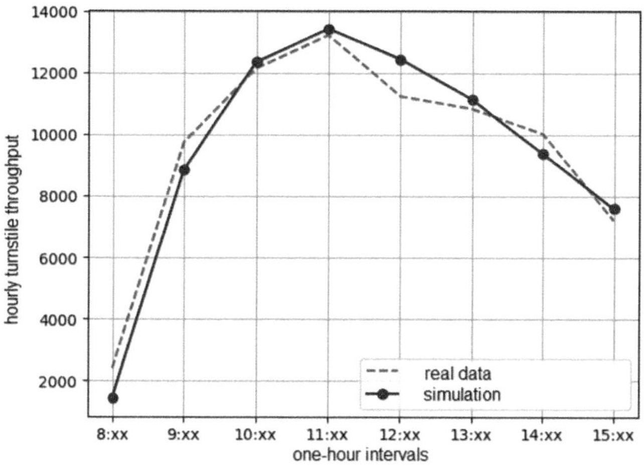

Fig. 9. Hourly arrivals at key turnstile (Real: dashed line, Simulated: solid line).

5 Conclusion

In summary, this paper delved into the efficacy of a MAS simulation in optimizing skier flow at a ski resort. By leveraging real-time data, the system aims to guide skier agents toward less congested slopes, thereby enhancing the overall skiing experience.

One of the primary findings of this research is the successful implementation of an AI-powered queue estimation system. By training an AI model on camera data, the system accurately estimates queue lengths at different slopes, allowing for the computation of real-time waiting times.

Moreover, the MAS integrates these waiting time estimates into a popularity-based decision making framework. By leveraging a probability density function derived from real data, the MAS calculates slope popularity metrics. Skier agents within the MAS utilize this information to make informed decisions, favoring slopes with shorter wait times.

Overall, the simulation demonstrates the potential of utilizing real-time data and AI techniques within a MAS framework to optimize skier flow. By incorporating queue estimates and popularity metrics, the system effectively guides skiers toward less crowded slopes, with the potential to reduce wait times and improve the skiing experience.

Future research endeavors can explore additional factors influencing skier decisions, such as individual skill levels or slope difficulties, to significantly enhance route recommendations. Integrating real-time weather conditions and considering dynamic pricing strategies within the MAS could provide even more comprehensive decision support for skiers. A promising avenue for future work involves developing a mobile application or real-time notification system that leverages the queue estimation system to inform skiers about wait times, empowering them to make data-driven decisions and optimize their time on the slopes.

Applying this framework to other leisure areas, such as theme parks, offers exciting opportunities to explore its versatility in managing crowds and enhancing visitor experiences.

Acknowledgments. We would like to thank Tatry mountain resorts, a.s. for providing the data, valuable suggestions, and additional support for this research. This work was supported by the grant KEGA 004UK-4/2024 "DICH: Digitalization of Cultural Heritage".

References

1. Barras, T., Doctor, M., Revilloud, M., Schumacher, M., Loubier, J.C.: Queues in ski resort graphs: the ski-optim model. In: 17th Agile Conference on Geographic Information Science, Castellón, June 2014
2. Delibašić, B., Radovanović, S., Jovanović, M., Vukićević, M., Suknović, M.: An investigation of human trajectories in ski resorts. In: Trajanov, D., Bakeva, V. (eds.) ICT Innovations 2017, pp. 130–139. Springer, Cham (2017)
3. Delibašić, B., Radovanović, S., Jovanović, M.Z., Suknović, M.: Improving decision-making in ski resorts by analysing ski lift transportation–a review. In: Mladenović, N., Sifaleras, A., Kuzmanović, M. (eds.) Advances in Operational Research in the Balkans, pp. 265–273. Springer International Publishing, Cham (2020)
4. Paszke, A., Gross, S., Massa, F., Lerer, A., Bradbury, J., Chanan, G., Killeen, T., Lin, Z., Gimelshein, N., Antiga, L., Desmaison, A., Kopf, A., Yang, E., DeVito, Z., Raison, M., Tejani, A., Chilamkurthy, S., Steiner, B., Fang, L., Bai, J., Chintala, S.: Pytorch: An imperative style, high-performance deep learning library. In: Wallach, H., Larochelle, H., Beygelzimer, A., d' Alché-Buc, F., Fox, E., Garnett, R. (eds.) Advances in Neural Information Processing Systems 32, pp. 8024–8035. Curran Associates, Inc. (2019), http://papers.neurips.cc/paper/9015-pytorch-an-imperative-style-high-performance-deep-learning-library.pdf
5. Poulhès, A., Mirial, P.: Dynaski, an agent-based model to simulate skiers in a ski area. Procedia Comput. Sci. **109**, 84–91 (2017), 8th International Conference on Ambient Systems, Networks and Technologies, ANT-2017 and the 7th International Conference on Sustainable Energy Information Technology, SEIT 2017, Madeira, Portugal, 16–19 May 2017
6. Poulhès, A., Mirial, P.: Modeling skier behavior for planning and management. dynaski, an agent-based in congested ski-areas (2020). https://doi.org/10.48550/arXiv.2011.11976
7. Pullman, M.E., Thompson, G.M.: Evaluating capacity- and demand-management decisions at a ski resort. Cornell Hotel Restaurant Admin. Quart. **43**(6), 25–36 (2002). https://www.sciencedirect.com/science/article/pii/S0010880402800696
8. Ren, S., He, K., Girshick, R., Sun, J.: Faster r-cnn: towards real-time object detection with region proposal networks. In: Advances in Neural Information Processing Systems, pp. 1468–1476 (2015)
9. Wang, B., Hess, V., Crooks, A.: Mesa-geo: a gis extension for the mesa agent-based modeling framework in python. In: Proceedings of the 5th ACM SIGSPATIAL International Workshop on GeoSpatial Simulation, GeoSim 2022, pp. 1-10. Association for Computing Machinery, New York, NY, USA (2022). https://doi.org/10.1145/3557989.3566157

Data Analysis for Cognitive States Identification

Snezhana Ribarska[1], Olga Georgieva[1(✉)], Sophia Lazarova[2], and for the Alzheimer's disease neuroimaging initiative

[1] FMI, Sofia University, "St. Kliment Ohridski", Sofia, Bulgaria
o.georgieva@fmi.uni-sofia.bg
[2] GATE Institute, Sofia University "St. Kliment Ohridski", Sofia, Bulgaria

Abstract. The current paper introduces an unsupervised data mining methodology applied to medical data to uncover insights into the cognitive states associated with Alzheimer's disease. The methodology emphasizes the integration of algorithmic feature selection with existing knowledge about the disease's manifestation. The data clustering analysis is conducted with two primary objectives: first, to uncover the underlying structure of the data, and second, to examine how well the identified cognitive groups align with existing diagnoses. The results demonstrate that the most effective clustering approach identifies three distinct cognitive groups. However, while each cluster represents a dominant cognitive group, it does not strongly differentiate between the subjects' diagnoses.

Keywords: Alzheimer's Disease · Unsupervised Data Mining · Genetic Algorithm · Cluster Analysis · FCM Clustering

1 Introduction

Dementia is a progressive and irreversible condition that adversely affects the brain function. Alzheimer's disease (AD) is the most common cause of dementia among older adults and contributes to about 60–70% of the cases [1]. Given that the onset of the disease is closely related to age and the fact that the overall life expectancy is growing, the cases of dementia are expected to increase dramatically over the next decades. According to the current predictions, over the next two decades the number of dementia cases is expected to triple and reach the staggering 150 million cases worldwide by 2050 [2]. Therefore, dementia and Alzheimer's disease are pressing societal matters that threaten to affect millions of people across the globe.

Machine learning has emerged as a powerful tool in Alzheimer's disease research, offering new insights into diagnosis, prognosis, and treatment strategies. By analyzing

Data used in preparation of this article were obtained from the Alzheimer's Disease Neuroimaging Initiative (ADNI) database (adni.loni.usc.edu). As such, the investigators within the ADNI contributed to the design and implementation of ADNI and provided data but did not participate in analysis or writing of this report. A complete listing of ADNI investigators can be found at: http://adni.loni.usc.edu/wp-content/uploads/how_to_apply/ADNI_Acknowledgement_List.pdf.

D. Simian and L. F. Stoica (Eds.): MDIS 2024, CCIS 2486, pp. 33–45, 2025.
https://doi.org/10.1007/978-3-031-87386-7_3

large datasets of medical imaging, genetic information, and clinical data, machine learning algorithms can identify patterns and biomarkers that may be indicative of early-stage Alzheimer's, often before symptoms are clinically evident [3–7].

While supervised machine learning methods are often applied in classification tasks, such as detection of Alzheimer's disease states [8], unsupervised analyses are advantageous in identifying clusters of natural cognitive groups based on proximity or similarity of the data. Thus, the class of unsupervised machine learning methods is a good candidate for modelling natural cognitive groups occurring on the AD spectrum. A systematic review [9] presents achievements of K-means, K-means mode, Hierarchal and Multi-Layer clustering to AD datasets. A study [10] employs clustering to identify distinct patterns of cognitive and functional decline in Alzheimer's patients. Other research combines statistical and unsupervised learning methods for analyzing magnetic resonance imaging (MRI) scans to detect AD-relevant regions for providing a basis for diagnosing and progressive mild cognitive impairment [11]. The efforts of [12] are directed to performance of a graphical neural network classifier based on data-driven diagnostic classes from unsupervised clustering on heterogeneous data. Feature reduction by t-SNE followed by K-means clustering was found to be the best distinguishing method among several ones performed on the ADNI dataset [13]. Then, association rules between medication usage, AD diagnosis, and AD subpopulation were determined. Finally, in our recent research an unsupervised data mining procedure is introduced for revealing cognitive groups of Alzheimer disease [14]. From the literature review, it can be concluded that the challenge of the large number and diversity of possible AD factors to consider complicates the machine learning procedure for deep understanding of the interdependence between diseases factors, appeared symptoms and related cognitive states. Solutions should be sought to improve the efficiency and accuracy of the applied data mining methodologies.

The current study complements this line of work by presenting an extended unsupervised data mining methodology for identification of AD cognitive groups. The goal of the work is to answer the following research question. How feature selection could take an advantage of different approaches both algorithmic and heuristic to increase the accuracy of the grouping? Two different approaches will be considered. First, introducing a new informative attribute to define the data space for improving the accuracy of the data partitions. Second, the data space is defined by features - biomarkers and the examinations of the cognitive tests measured in different time in order to reveal the cognitive groups more precisely by reflecting the biological nature of the disease. Further, cluster analysis tolerant to uncertainty is applied for identifying cognitive groups and to assess how they are covered by the existing cognitive diagnosis. Interpretation of the results retrieved from the groups found is helpful for disease planning and tailoring the patients' treatment.

2 Data

The data used in the preparation of the present work was obtained from the Alzheimer's disease Neuroimaging Initiative (ADNI) database. The ADNI was launched in 2003 as a public-private partnership, led by Principal Investigator Michael W. Weiner, MD.

The primary goal of ADNI has been to identify and validate biomarkers for the early detection and tracking of Alzheimer's disease [15]. For up-to-date information about ADNI, see adni.loni.usc.edu.

3 Data Mining Methodology

According to the purpose of the study, the real structure of the data should be revealed in order to distinguish the groups of a certain cognitive state. A specifically elaborated unsupervised data mining procedure is proposed in Fig. 1. It is described in detail in the next subsections.

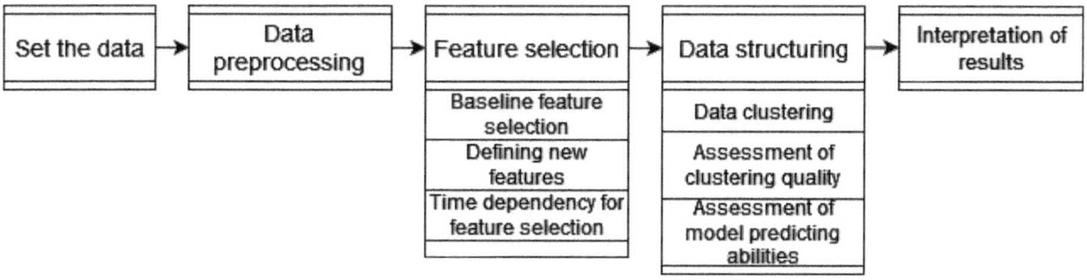

Fig. 1. Stages of the data mining methodology.

3.1 Data Set for Cognitive State Retrieval

As ADNI offers large volumes and variety of data, we looked to obtain some subset that is in line with the current understanding of Alzheimer's disease biomarkers, symptoms and pathophysiology. Thus, the dataset obtained for the preparation of the present work was structured as follows.

Diagnosis. All ADNI participants were placed in one of the following groups – cognitively normal (CN), significant memory concern (SMC), early mild cognitive impairment (EMCI), late mild cognitive impairment (LMCI) and Alzheimer's disease (AD). Individuals in the SMC group are cognitively normal individuals that have expressed concerns about their memory. Therefore, the SMC group is essentially a subgroup of CN. Similarly, EMCI and LMCI are subgroups of mild cognitive impairment (MCI), the first clinical stage of Alzheimer's disease. Detailed inclusion/exclusion criteria is available in the official ADNI documentation [16]. It should be underlined that the applied unsupervised approach does not use the diagnosis labels for grouping processing but only to estimate the retrieved data structure and for results interpretation.

Demographic Data. Demographic data included age (AGE), gender (PTGENDER), years of education (PTEDUCAT) and body mass index (BMI) estimation based on weight and height.

Biomarkers. The obtained data contained assessments of various biomarkers, including cerebrospinal fluid (CSF) biomarkers, imaging and genetic biomarkers. The neuropathological markers of AD were represented by measurements of the concentrations of amyloid beta 42 (Aβ42) and phosphorylated tau (p-Tau) in CSF. CSF measurements of total tau was also included as a proxy of neurodegeneration. Additionally, data from florbetapir positron emission tomography (AV45-PET) was also included as an alternative measure of amyloid burden. Measures of global and local cortical atrophy were included as global cortical volume (WholeBrain_NORM) and hippocampal volumes (Hippocampus_NORM), measured with magnetic resonance imaging. Hippocampal volumes were selected as a proxy of medial temporal atrophy, a neurodegenerative process characteristic for Alzheimer's disease. Measures of glucose metabolism as observed with fludeoxyglucose-18 (FDG) positron emission tomography (FDG-PET) were also included.

Cognition and daily function. Several measurements of cognition by cognitive tests were included in the dataset, namely Mini-Mental State Examination (MMSE) [17], Alzheimer's disease Assessment Scale Cognitive (ADAS13) [18]. Additionally, the severity of dementia was indicated by the Clinical Dementia Rating scale (CDRSB) [19] and the ability to handle various activities of daily living was assessed with the Functional Activities Questionnaire (FAQ) [20]. The final dataset included the total score for each of the upper-mentioned assessments. Finally, the total number of correctly generated items in a category-naming task (animals) was included as a measure of verbal fluency (CATANIMSC), a cognitive function that refers to the ability to generate words quickly and efficiently within certain constraints. Poor performance on verbal fluency tasks can indicate issues with cognitive processing speed, language production, or executive control, and is often associated with various forms of cognitive impairment, including Alzheimer's disease [21, 22].

3.2 Data Preprocessing

Some problems with raw data set should be solved prior the actual work of analysis. Participants who were not examined at some of their visits or at some examinations due to reasons such as absence, difficulty in performing or taking a lot of time for the examination were filtered out.

A part of the data is nominal. Thus, demographic information for gender and education as well as the diagnosis data were transformed into numeric values.

The problem of different data scaling is solved by max-min normalization. It is an initial obligatory step of the clustering algorithm that prevents distortion of the result. Also, this transformation can improve the readability and interpretation of the results.

3.3 Feature Selection

The proposed extended feature selection solution has three stages. In the beginning the selection was done on baseline data that were collected at the first participant visit. Then, the number and type of the selected features was refined in two directions. First, a new feature was calculated by some selected attributes. Second, the selection was improved regarding the appropriate choice of features' values in time.

The task for feature selection of the baseline data was solved by Genetic algorithm. It is an optimization algorithm inspired by the process of natural evolution able to deal with large volumes of data. It has advantages of a robust method that can detect global extremes [23, 24]. The algorithm is a variant of stochastic search where new states are generated by combining pairs of parent states. States are represented as strings over a given finite alphabet formed by the investigated features. The evaluation function, named also fitness function, estimates the closeness to the goal of each defined state. The fitness function has larger values for better states. Linear regression is commonly accepted for fitness function as it enables to measure the correlation between variables and capable of handling noise in the data. The feature selection is done according to the number of occurrences of each feature to maximizing the value of the fitness function.

Defining New Informative Attribute. A new informative attribute that defines the data space could improve the accuracy of the data partition. The inspiration of such idea is the medical practice of Alzheimer's diagnosis. The disease diagnosis is based not on one but on a set of cognitive examinations. We suggest that all cognitive features that have been selected in the feature selection step to be replaced by one feature that accounts for the contribution of each selected feature. For this aim we propose the following calculation formula for the new feature F:

$$F = \frac{a_1F_1 + a_2F_2 + \cdots + a_nF_n}{n}, \tag{1}$$

where $F_1, F_2, \ldots, F_n$ are features of n cognitive tests that have been selected by the feature selection algorithm. The coefficients $a_1, a_2, \ldots a_n$ are weighted coefficients estimated as a ratio of the number of occurrences of each feature found by the Genetic algorithm to the total number of the occurrences of all features F_i, $i = 1, \ldots, n$. The calculated new feature F is accepted as a dimension of the data space instead of all constituent features F_i, $i = 1, \ldots, n$. The effect of this is double - reducing the data dimensionality, which facilitates the subsequent data structuring. Also, the estimation (1) ensures that the cognitive tests participate in the final dependency model according to their level of importance.

Time Dependency Account for Feature Selection. It is well-known that the Alzheimer's diagnosis based on cognitive tests in fact is a late estimation of the cognitive impairment as the disease is manifested years later after than the occurrence of the brain disability. On the other hand, the biomarkers could give an assessment of the current brain impairment. Having this knowledge, it could suppose that by forming a data space defined by the biomarkers and the examinations of the cognitive tests done at later time could reveal the cognitive groups more precisely. By following this aim we have organized a new data set by biomarkers' data of the first visit combined with next visits cognitive tests results. This data set was processed by the proposed data mining methodology.

3.4 Data Clustering

Data clustering is applied to find the natural grouping of subjects with a certain cognitive status. The data spaces defined in the Feature selection stage could be further compared according to their distinguishing abilities. The right choice of the applied clustering algorithm is important as it needs to deal with the existing uncertainty. The disease develops slowly and the brain decline occurs gradually, which makes separation difficult. Personal characteristics also influence the development of dementia.

Soft computing clustering algorithms can deal with this problem as they describe the lack of a clear boundary between groups. The most used algorithm that covers this demand is Fuzzy C-means (FCM) [25]. It assigns each data to each cluster with a degree of membership between 0 and 1. FCM is an objective function algorithm with clustering criteria J that minimizes the sum:

$$J = \sum_{i=1}^{c} \sum_{k=1}^{N} u_{ik}^{m} d_{ik}^{2}, \tag{2}$$

where u_{ik} denotes the membership degree of the data point x_k, $k = 1, \ldots, N$, to the i-th cluster center v_i, $i = 1, \ldots, c$; N is the number of the data points and c is the number of clusters. The coefficient m determines how much clusters may overlap. Euclidean distance is applied for distance measure $d_{ik} = \|x_k - v_i\|$. The number of clusters c has to be set in advance. By applying alternating optimization scheme, FCM finds the cluster centers v_i and the membership degrees u_{ik}.

Assessment of the Clustering Quality. The assessment of the grouping is necessary to measure the right number of clusters for each data space. The quality of grouping admits also to compare the data spaces between them as better partitioning is more informative. The clustering quality is estimated by cluster indexes [26]. Here we apply two of them. Average Within-Cluster Distance (*AWCD*) estimates how compact and how separated clusters are:

$$AWCD = \frac{1}{c} \sum_{i=1}^{c} \frac{\sum_{k=1}^{N} u_{ik}^{m} * d_{ik}^{2}}{\sum_{k=1}^{N} u_{ik}^{m}}. \tag{3}$$

The “knee” (or “elbow”) heuristic mathematical method is applied for choosing the right number of clusters. This method consists of presenting the *AWCD* value against the number of clusters c obtained by the clustering of a certain data space. The knee/elbow of this dependency shows the best number of clusters. This is a point where the increasing the number of clusters is no longer worth the additional *AWCD* decreasing.

The Partition entropy (*PE*) describes fuzzy uncertainty of the partition as follows:

$$PE = -\frac{1}{N} \sum_{k=1}^{N} \sum_{i=1}^{c} u_{ik} * \log(u_{ik}), \tag{4}$$

which varies between 0 and $\log(c)$. The lower *PE* value shows better clustering.

Assessment of Model Predicting Abilities. The other interesting question of the clustering quality is how far the clusters found overlap the given diagnoses. For each clustering the statistical distribution of the participants’ diagnoses in each cluster are calculated. Since the methodology applies fuzzy clustering it is assumed that a subject belongs to that cluster to which the highest membership degree is obtained.

3.5 Interpretation of the Data Mining Results

Interpretation of results is possible in several directions. First, the extracted significant features with their level of significance conclude about the features that are important for diagnosis. Second, the found clusters presents the natural structure of the data space and ability to form cognitive groups. Third, it is important to answer to what extend the information of the real data structure coincides with the diagnosed cognitive status. Defined cluster centers are a source of information as their coordinates could be interpreted according to their respective referent medical values of the attributes examined.

4 Results and Discussion

From the total amount of 2405 available data points for cognitive state retrieval after the preprocessing described data of 1541 participants remains for the next data mining steps.

Python 3 was used for the implementation of the data mining procedure with the following libraries used: NumPy, CSV, Skfuzzy, Sklearn, Matplotlib.

4.1 Feature Extraction Results

The Genetic algorithm takes as input the preprocessed data of the participants examined at first visit. The first two hundred best presented individuals in terms of the fitness function are taken in the selection process. Experiments show that by increasing their number the selection results are not changed, however the execution time of the algorithm increases significantly. Every two of these individuals are grouped in pairs on a random principle. The crossover between the pairs was performed at one random point. Further the experiments are done by varying a parameter K that stands for the number of independent runs of the algorithm.

The more a feature occurs, the more informative this feature is to the dependency model. It could be seen (Table 1) that PTAU and ABETA features are presented most and in each trial experiment. It is important to note that AGE feature also has quite a large dependency impact. Among the cognitive tests the MMSE is best presented and is one of the most sensitive attributes. However, other tests CDRSB, FAQ, ADAS13 also have certain impact. The education feature PTEDUCAT is questionable as nevertheless it shows good appearance this variable has no such diversity values, which will bother the next clustering analysis.

The Body Mass Index is occurred at large, but not with the same frequency as the already mentioned features. However, it should be underlined that the relationship between BMI and Alzheimer disease is complex and not directly causal. Factors such as genetics, lifestyle, overall health, and disease progression, can also affect person's BMI. The rest of the features of Table 1 are not found as significant.

In seeking a more accurate dependency model a new attribute that generalizes the cognitive assessment features was defined. According to the explanations about the new attribute, F was calculated by formula (1). The constituting features F_i, $i = 1,..,3$ of Eq. (1) are the cognitive tests that had been recognized as important for the model

Table 1. Number of occurrences of each feature estimated by the Genetic algorithm.

K	AGE	PTEDUCAT	MMSE	ADAS13	CDRSB	FAQ	FDG	AV45	ABETA	PTAU	TAU	BMI	CATANIMSC	Hippocampus_NORM	WholeBrain_NORM	PTGENDER
5	4	2	1	0	1	0	0	0	5	5	0	0	0	0	0	0
10	7	0	5	0	4	4	0	0	6	7	3	4	0	0	0	0
13	7	5	7	0	0	0	0	0	9	10	0	5	0	0	0	0
20	13	4	11	1	2	0	0	0	13	12	11	10	0	0	0	0

(Table 1). They are respectively MMSE, FAQ, ADAS13. Nevertheless, that the CDRSB test also was detected as impact feature it was not included in forming the attribute *F* due to the prescribed medical recommendation to use it alone in disease assessment [16].

In the further experiments beside the biomarkers' values at the first visit the composite attribute *F* estimated of a subsequent visit was investigated. These experiments try to describe the real situation as the disease visibility at the person's behavior is later than the impairment processes in the brain. Since most patients had an examination at 12 months visit, it is representatively to explore the informativity of a data space combining the baseline biomarkers' values and these late cognitive assessments.

4.2 Clustering Analysis

The next experiments and comparison analysis underline the ability of the proposed approach for distinguishing the participants' groups. Three distinct experiments in regard the formed data spaces were conducted. The most significant attributes that are the two biomarkers ABETA and PTAU are included to form the three data spaces but differ in other constituent attributes. For the first experiment E1 the data space is added with the MMSE data and with AGE in a particular experiment. For the second experiment E2 the new attribute *F* instead MMSE is included. The third experiment E3 also takes the common attribute *F* but calculated for the data at the 12th month visit noted as *F_m12*.

Further, FCM clustering is applied for default values of the parameter $m = 2$ and the accuracy $tol = 0.005$. By varying the number of clusters from two to six, the partition abilities of the data spaces are compared in terms of the clustering indexes given by Eqs. (3) and (4) to find the best data structuring.

All experiments show that the best partitioning is found for clustering in three clusters, $c = 3$, where the "knee" of *AWCD* curve and the minimum of *PE* are found (Table 2). From the two data spaces of E1 the preference is given to the space [ABETA, PTAU, MMSE] as $PE = 0.12$ has lower value. The comparison of the three experiments clearly shows that the generalized cognitive attribute *F* improves the partitioning. Best data structuring is obtained for the data space E3 with a common attribute of the 12-month examinations, *F_m12* and for which the Partition entropy has the lowest value, $PE = 0.06$.

Certain source of information are the clusters centers of the best partitions (Table 3), where it could be seen that the three cognitive states are well distinguished. Amyloid beta

Table 2. Cluster indexes of the grouping in different data spaces.

	Cluster index					
	c	2	3	4	5	6
E1	ABETA, PTAU, MMSE, AGE					
	AWCD	5479.16	2593.09	1568.11	1354.70	1011.45
	PE	0.32	0.15	0.43	0.47	0.62
	ABETA, PTAU, MMSE					
	AWCD	5112.37	3924.88	3783.78	2040.88	822.62
	PE	0.31	0.12	0.36	0.46	0.40
E2	ABETA, PTAU, *F*					
	AWCD	7784.29	3266.46	3097.81	2904.14	2717.12
	PE	0.41	0.08	0.36	0.46	0.38
E3	ABETA, PTAU, *F_m12*					
	AWCD	5980.12	2912.43	967.09	803.65	786.64
	PE	0.29	0.06	0.23	0.40	0.52

concentrations in CSF below 1000 indicate the presence of pathological accumulations in the brain [16]. Similarly, phosphorylated tau (PTAU) concentrations above 27 pg/ml are considered for a positive reading and presence of tau pathology [16]. MMSE results between 20–26 indicate dementia, which also strengthens the clustering result for the case of experiment E1.

Table 3. Cluster centers of partitioning in three clusters, $c = 3$.

	E1			E2			E3		
№	ABETA	PTAU	MMSE	ABETA	PTAU	*F*	ABETA	PTAU	*F_m12*
1	749.97	24.89	27.51	746.35	24.88	0.63	1525.20	21.74	0.61
2	1573.35	20.76	28.89	622.84	36.23	2.50	709.25	28.71	1.17
3	530.51	35.72	22.86	1603.14	20.70	0.47	573.82	36.33	4.04

Other important question is to what extend the groups cover the medical diagnoses. The space E1 is valuable (Fig. 2) as Cluster 1 presents at large the MCI state as a sum of 31.5% LMCI and 26.1% EMCI diagnoses. Largest group of Cluster 2 is the healthy people formed by CN = 36.1% and SMC = 21%. Cluster 3 represents AD diagnosed of 60.2%. Cluster 1 of E2 data space (Fig. 3) represents at most MCI as a sum of EMCI 25.5% and 31.3% LMCI, Cluster 2 - AD participants of 69.6%, whereas Cluster 3 is valuable for healthy people representing 36.9% CN and 21.2% SMC. The E3 space does not account for five diagnosis because these are assessed only at the first visit (Fig. 4).

Cluster 1 recognizes healthy participants 56.7% CN, Cluster 2–64.1% MCI and Cluster 3–56.2% AD.

Each cluster presents a leading cognitive group but partitions do not strongly distinguish the subjects in their diagnoses. Partly this result is due to the lack of well-established disease markers especially in the early stages, when the diagnosis could not be well done. However, these results confirm the objectivity that there is no a strict boundary between groups. In this sense the clustering membership degree obtained through the applied (

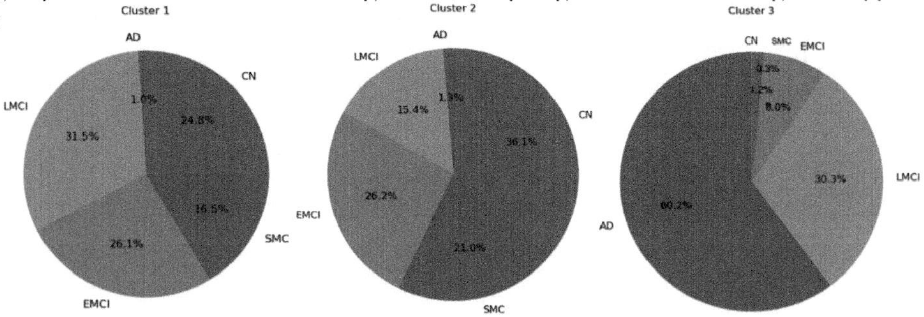

Fig. 2. Percentage of diagnoses in the three clusters of the space ABETA, PTAU, MMSE.

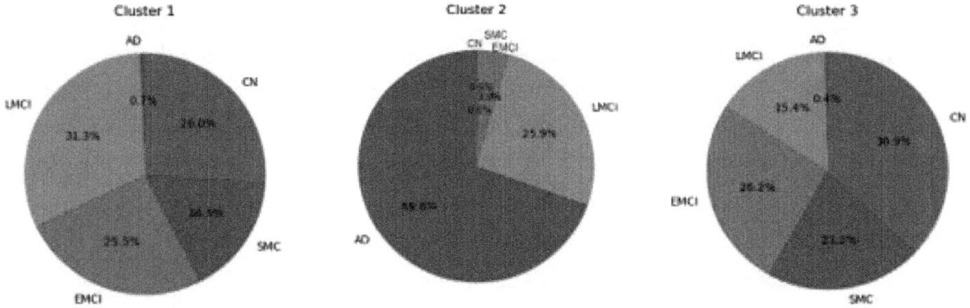

Fig. 3. Percentage of diagnoses represented in the space ABETA, PTAU, *F*.

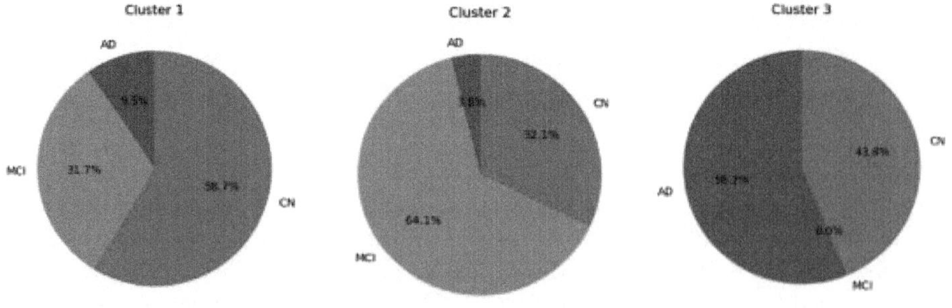

Fig. 4. Percentage of diagnoses represented in the space ABETA, PTAU, *F_m12*.

5 Conclusion

This paper presented a data mining methodology designed to uncover insights into the cognitive states of Alzheimer's disease through the analysis of medical data. The methodology emphasizes the critical importance of feature selection, as it aims to identify the most significant attributes from a large pool of existing features while avoiding those that are costly and difficult to examine. To achieve this, a Genetic Algorithm was employed. Its results were further enhanced with specific heuristic extensions.

The study identified amyloid beta and phosphorylated tau as the most significant factors, followed by the cognitive test MMSE. The capability of a new compound feature to generalize over measurements of cognition and improve differentiation between cognitive states was explored. The results indicated that the inclusion of a compound feature instead several cognition examinations improves the data partitioning, in terms of cluster indexes. The data partitioning results are improved if the data space is defined by features of biomarkers and the examinations of the cognitive tests measured in late time. Thus, it reveals the cognitive groups more precisely covering the disease manifestation.

The presented unsupervised data mining approach served for two purposes: first, to uncover the underlying data structure, and second, to determine how well diagnoses are represented within the identified cognitive groups. The findings demonstrate that the optimal clustering solution reveals three distinct groups across all of the investigated data spaces. The results also indicated that while each identified cluster represents a dominant cognitive group, it does not strongly differentiate between the subjects' diagnoses. The interpretation of the results retrieved from the groups found helps for disease planning and tailoring the patient's treatment. They are important for feature development of effective early stage Alzheimer's disease diagnosis tool.

Acknowledgments. Acknowledgments. This research work has been supported by GATE project, funded by the Horizon 2020 WIDESPREAD-2018-2020 TEAMING Phase 2 program under grant agreement no.857155 and by the European Union-NextGenerationEU, through the National Recovery and Resilience Plan of the Republic of Bulgaria, project SUMMIT BG-RRP-2.004-0008-C01 and funded by Science Fund of Sofia University by project no. 80-10-137/2024. Data collection and sharing for this project was funded by the Alzheimer's Disease Neuroimaging Initiative (ADNI) (National Institutes of Health Grant U01 AG024904) and DOD ADNI (Department of Defense award number W81XWH-12-2-0012). ADNI is funded by the National Institute on Aging, the National Institute of Biomedical Imaging and Bioengineering, and through generous contributions from the following: AbbVie, Alzheimer's Association; Alzheimer's Drug Discovery Foundation; Araclon Biotech; BioClinica, Inc.; Biogen; Bristol-Myers Squibb Company; CereSpir, Inc.; Cogstate; Eisai Inc.; Elan Pharmaceuticals, Inc.; Eli Lilly and Company; EuroImmun; F. Hoffmann-La Roche Ltd and its affiliated company Genentech, Inc.; Fujirebio; GE Healthcare; IXICO Ltd.; Janssen Alzheimer Immunotherapy Research & Development, LLC.; Johnson & Johnson Pharmaceutical Research & Development LLC.; Lumosity; Lundbeck; Merck & Co., Inc.; Meso Scale Diagnostics, LLC.; NeuroRx Research; Neurotrack Technologies; Novartis Pharmaceuticals Corporation; Pfizer Inc.; Piramal Imaging; Servier; Takeda Pharmaceutical Company; and Transition Therapeutics. The Canadian Institutes of Health Research is providing funds to support ADNI clinical sites in Canada. Private sector contributions are facilitated by the Foundation for the National Institutes of Health (www.fnih.org). The grantee organization is the Northern California Institute for Research and Education, and the study is coordinated by the

Alzheimer's Therapeutic Research Institute at the University of Southern California. ADNI data are disseminated by the Laboratory for Neuro Imaging at the University of Southern California.

Disclosure of Interests. The authors have no competing interests to declare that are relevant to the content of this article.

References

1. World Health Organization (WHO), Dementia Factsheet, https://www.who.int/news-room/fact-sheets/detail/dementia, Accessed 27 Sep 2024
2. Nichols, E., et al.: Estimation of the global prevalence of dementia in 2019 and forecasted prevalence in 2050: an analysis for the global burden of disease study 2019. Lancet Public Health **7**(2), e105–e125 (2022)
3. Alatrany, A.S., Khan, W., Hussain, A., Kolivand, H., Al-Jumeily, D.: An explain-able machine learning approach for Alzheimer's disease classification. Sci. Rep. **14**(1), 2637 (2024)
4. Ding, Y., et al.: A deep learning model to predict a diagnosis of Alzheimer disease by using 18 F-FDG PET of the brain. Radiology **290**(2), 456–464 (2019)
5. Diogo, V.S., Ferreira, H.A., Prata, D.: For the Alzheimer's disease neuroimaging initiative: early diagnosis of Alzheimer's disease using machine learning: a multi-diagnostic, generalizable approach. Alzheimer's Res. Therapy **14**(1), 107 (2022)
6. Kavitha, C., Mani, V., Srividhya, S.R., Khalaf, O.I., Tavera Romero, C.A.: Early-stage Alzheimer's disease prediction using machine learning models. Front. Public Health **10**, 853294 (2022)
7. Koychev, I., Marinov, E., Young, S., Lazarova, S., Grigorova, D., Palejev, D.: Identifiction of preclinical dementia according to ATN classification for stratified trial recruitment: a machine learning approach. PLoS ONE **18**(10), e0288039 (2023)
8. Franciotti, R., Nardini, D., Russo, M., Onofrj, M., Sensi, S.L.: Comparison of machine learning-based approaches to predict the conversion to Alzheimer's disease from mild cognitive impairment. Neuroscience **514**, 143–152 (2023)
9. Alashwal, H., Halaby, El., M., Crouse, J.J., Abdalla, A., Moustafa, A.A.: The application of unsupervised clustering methods to Alzheimer's disease. Front. Comput. Neurosci. **13**, 31 (2019)
10. Geifman, N., Kennedy, R.E., Schneider, L.S., Buchan, I., Brinton, R.D.: Data-driven identification of endophenotypes of Alzheimer's disease progression: implications for clinical trials and therapeutic interventions. Alz Res. Therapy **10**(1), 4 (2018)
11. Liu, Y., Mazumdar, S., Bath, P.A.: An unsupervised learning approach to diagnosing Alzheimer's disease using brain magnetic resonance imaging scans. Int. J. Med. Inf. **173**, 105027 (2023)
12. McCombe, N., et al.: Alzheimer's disease classification using cluster-based labelling for graph neural network on heterogeneous data. Healthcare Tech. Lett. **9**(6), 102–109 (2022)
13. Prakash, J., Wang, V., Quinn, R.E., Mitchell, C.S.: Unsupervised machine learning to identify separable clinical Alzheimer's disease sub-populations. Brain Sci. **11**(8), 977 (2021)
14. Ribarska, S., Georgieva, O.: Unsupervised analysis of Alzheimer disease medical data. In: Proceedings of the 2023 6th International Conference on Computational Intelligence and Intelligent Systems, pp. 32–38. ACM, Tokyo, Japan (2023)
15. Weiner, M.W., Aisen, P.S., Jack, C.R.Jr, Jagust, W.J., et al.: Alzheimer's disease neuroimaging initiative. The Alzheimer's disease neuroimaging initiative: progress report and future plans. Alzheimers Dement **6**(3) 202–11.e7 (2010)

16. Alzheimer's disease neuroimaging initiative: ADNI 3 procedures manual, Accessed 12 Apr 2024. https://adni.loni.usc.edu/wp-content/uploads/2012/10/ADNI3-Procedures-Manual_v3.0_20170627.pdf
17. Folstein, M.F., Folstein, S.E., McHugh, P.R.: Mini-mental state. J. Psychiatr. Res. **12**(3), 189–198 (1975)
18. Rosen, W.G., Mohs, R.C., Davis, K.: L: A new rating scale for Alzheimer's disease. AJP **141**(11), 1356–1364 (1984)
19. Morris, J.C.: The clinical dementia rating (CDR): current version and scoring rules. Neurology **43**(11), 2412 (1993)
20. Pfeffer, R.I., Kurosaki, T.T., Harrah, C.H., Chance, J.M., Filos, S.: Measurement of functional activities in older adults in the community. J. Gerontol. **37**(3), 323–329 (1982)
21. Saunders, N.L.J., Summers, M.J.: Attention and working memory deficits in mild cognitive impairment. J. Clin. Exp. Neuropsychol. **32**(4), 350–357 (2010)
22. Henry, J.D., Crawford, J.R.: A meta-analytic review of verbal fluency performance following focal cortical lesions. Neuropsychology **18**(2), 284–295 (2004)
23. Kohavi, R., John, G.H.: Wrappers for feature subset selection. Artif. Intell. **97**(1–2), 273–324 (1997)
24. Han, J., Kamber, M., Pei, J.: Data mining: concepts and techniques, 3rd edn. Elsevier, USA (2012)
25. Bezdek, J.C.: Pattern Recognition with Fuzzy Objective Function Algorithms. Boston MA (1981)
26. Eustáquio, F., Nogueira, T.: Evaluating the numerical instability in fuzzy clustering validation of high-dimensional data. Theoret. Comput. Sci. Comput. Sci. **805**, 19–36 (2020)

Evolutionary Computing and Metaheuristics

Ant Colony Optimization for Planting Vegetables and Annual Fruit Crops

Stefka Fidanova[1(✉)], Ivan Dimov[1], and Denitsa Angelova[2]

[1] Institute of Information and Communication Technologies, Bulgarian Academy of Sciences, Sofia, Bulgaria
stefka.fidanova@gmail.com

[2] Euro-Mediterranean Center on Climate Change, Ca' Foscari University of Venice, Venice, Italy
denitsa.angelova@cmcc.it

Abstract. Growing vegetables and annual fruit crops is an important part of agriculture. They are irreplaceable part of the human diet and are especially important for the body's supply of vital vitamins and minerals. Planning the areas for sowing is a complex task. Expected yields, different types of costs, policies, incentives, and more must be considered. The aim is for farmers to plan their sowing in such a way that they have maximum profit. On the other hand, authorities should be able to assess in advance the impact of various policies, including taxes and incentives, on farmers' sowing decisions. The problem is a complex optimization problem because of the strict constraints that must be satisfied. For problem solving, we apply an algorithm based on the ant colony optimization metaheuristics. In the study, we play out a variety of seeding scenarios involving a distinct opportunity for an adverse event occurring as well as subsidizing unattractive plantings.

Keywords: Agricultural system modelling · Planting vegetables · Ant colony optimization · Metaheuristics

1 Introduction

The production of fruits and vegetables is an important subsector of agriculture and, in particular, crop production. They are of great importance as fresh food for humans and a source of vitamins and minerals. They are also of great importance for the food and canning industry. The production of vegetables and annual fruit crops can be done outdoors in vegetable gardens as well as in greenhouses. Each type of production has its specifics. Outdoor production is affected by weather, soils, weather conditions, but have better taste and nutritional qualities. Vegetables and annual fruit crops need more water than cereals. A drought can lead not only to a reduction in the harvest, but even to its failure. Therefore, where they are sown, there must be an opportunity for irrigation. In order to grow vegetables successfully, many factors must be considered.

D. Simian and L. F. Stoica (Eds.): MDIS 2024, CCIS 2486, pp. 49–59, 2025.
https://doi.org/10.1007/978-3-031-87386-7_4

In the cultivation of cereals, all operations are mechanized, from tillage to harvesting. Vegetables also require manual labor, especially during harvesting. Some vegetables require too much manual labor, which makes them more expensive. A large part of the vegetables are perishable and require special conditions for storage and transportation. This makes their cultivation radically different from that of cereals. Growing vegetables has more costs than growing grains, but on the other hand, vegetables have a higher selling price that offsets the costs. This makes them attractive to grow. Using a variety of incentives and policies, authorities can also make vegetable growing attractive.

These circumstances are accepted by the policies of the European Union (EU). The EU seeks an optimization of the use of scarce natural resources such as water and land especially considering climate change. Using organic farming by reducing the use of chemicals and soil improvers [26]. The Sixth Report (AR6) of the International Panel for Climate Changes (IPCC), published in August 2021, states that "Each of the last four decades has been successively warmer than any decade that preceded it" (Paragraph A.1.2., page 6 in [17]). Concentration of pollutants, which can make harmful effect on plants, depends on temperature and their effect can be increased in increase of temperature. Therefore it is important to study the effects of air pollution on plants.

Agricultural production depends on farms, so appropriate incentives and policies are particularly important for its development. This is an incentive to optimize agricultural production. By creating a variety of models, the possibilities for optimization and increasing efficiency can be tested. The review work [21] is focused on bio-economic farm models. Cotton production model is proposed in [5]. Managering flexibility is shown in [24] by implementing stochastic optimization model. Combination between investment and management grips were used in [25]. The multi objective model proposed in [23] is for producing wood biomass. Some comparison of models is presented in [3].

In [12] is proposed calculation of nitrogen impact of agricultural production. Changes of the climate and its influence on some agricultural crops is discussed in [11,13,14,19]. In [15] are simulated EU farming system and politics. The importance of long-term prediction of change of the climate is discussed in [16]. In [20,22] uncertainty are included. The impact of ozone pollution on rice production is reviewed in [1]. Estimation the effect of intercropped leguminous service crops on main crops and soil is made in [6].

The existing models for agricultural production refer mainly to the cultivation of cereals or technical crops. They do not consider the probability of an adverse event occurring. Models for the production of vegetables and annual fruit crops are rare, therefore, in this study we focused on planting vegetables and annual fruit crops. The definition of the problem is made as an optimization problem. The result shows for each crop on what part of the available land it will be sown. The goal is for the farmer to get maximum income, taking into account all costs and events that may affect it including uncertainties. The model was made using ant colony optimization (ACO) [4].

We have developed test cases with a variety of scenarios. First, the model was tested under normal conditions without including subsidies, additional treatments and the probability of adverse events. A drought probability was then added to the model testing. A variant with a low probability of drought and a variant with a high probability of drought were tested. With a high probability of drought, additional watering of the most affected crop was added. This leads to an increase in the harvest, but also to additional costs. The impact of subsidies and how large they should be to make a less profitable crop attractive to farmers is explored.

The paper organization is as follows. Discussed problem is defined in Sect. 2. ACO algorithm is proposed in Sect. 3. In Sect. 4 some tests and discussion of the results are done. Section 5 is dedicated to concluding remarks.

2 Definition of the Problem

Vegetables are a large group of foods that include various parts of edible plants. Thus, vegetables most often refer to the leaves, stems, or roots of certain plants, but also fruits with lower sugar content, such as tomatoes, or seeds. For millennia, low agricultural productivity has limited the diversity of food for the majority of the population. Adding vegetables is important to diversifying such a diet. Growing and eating vegetables and fruits are of particular importance for proper human nutrition. They are the main supplier of vitamins and minerals for the human body. Regardless of the fact that these substances are in small quantities, they are extremely important for people's health.

2.1 Problem Description

The modeling problem of growing vegetables and annual fruit crops can be defined as a farmers' profit maximization problem. The basic assumption is that farmers want to increase their profit from sowing vegetables. To achieve this they have to decide which crops to sow on which plots and in what quantity. To make this decision, the farmer takes into account a variety of conditions: expected yield; land costs; plant cultivation costs; subsidy and incentive opportunities; likelihood of additional costs or crop reductions due to pests or adverse weather conditions; cost of production. The problem is to obtain the maximum profit when there are several plots to plant and the availability of several crops that could be planted. The size of the plots they have and which plants they are more suitable for is also of great importance. Irrigation opportunities should not be overlooked. During droughts, plots with good irrigation opportunities may be more suitable than other plots that are more fertile under normal weather conditions. As a solution to the problem, the farmer receives a recommendation on which plot to sow which crops and in what quantity. A minimum number of crops to sow was introduced to prevent only one crop, the most profitable, being sown. For each crop, a minimum sowing area is fixed for the cultivation of the crop to be efficient.

2.2 Model Input Data

The decision making farm model is implemented in the computer language C, and the input data are:

- available plots and their size;
- crops for sowing;
- minimum seeding area for each plan;
- cost of production;
- unit area subsidy for individual plantations;
- expected yield unit area yield of each plot;
- cultivation costs (processing, planting, treatment, feeding, etc.)
- land costs (fees, rents, etc.)
- probability of an adverse event (drought, air pollution, pests, etc.);
- reduction coefficient of yield in case of adverse event.

Seeding and cultivation data for squash, pumpkins and watermelons were used to test the model. These vegetable and fruit crops were selected because they are widely used on the territory of Bulgaria. On the other hand, they have similar soil requirements and can be sown on the same terrain together. The tillage of these crops is also similar. We use data from Bulgarian Chamber of Agriculture for calibrating the model. The data used are for northeastern Bulgaria for the year 2022. The land in Bulgaria is traditionally measured in acre (1 acre = 0.1 hectare). All costs are measured in Euro.

3 Ant Colony Optimization

Nature can give us many examples of optimal behavior. Such examples are the search for food, migration processes, hunting, survival actions, etc. When working in a group, good coordination is of utmost importance to the success of the group. This is the so-called group intelligence. Animals in which the individual does not show a high level of intelligence can show a high level of group intelligence. Examples include bees or ants. Scientists have long noticed intelligence in the group behavior of animals. These observations of theirs have given them ideas for new optimization methods.

The Ant Colony Optimization (ACO) method is between the best methods for solving discrete optimization problems. It is a stochastic method belonging to the set of metaheuristic methods. Ants in the nature deposit on their path an aromatic substance called pheromone. If there is no pheromone in the area where they are looking for food, they move randomly. When there are pheromone trails, they follow the most concentrated pheromone. This way of searching helps them track down the shortest path between two objects. The pheromone is reduced naturally by evaporation. Thus the less used traces gradually become unattractive.

3.1 Main ACO

Computationally complex problems cannot be solved efficiently by exact or traditional numerical methods. They are approached with metaheuristic methods. Their flexibility helps solve complex problems [4,7–10,18].

The first application of an algorithm mimicking ants in nature was made by Marco Dorigo [2]. The basic of the ACO methodology is the graph representation of the problem and simulation of searching a food by ants. Representation of the solutions is by paths in a graph and the purpose is to look for a shorter path according to given constraints. Main difference of ant colony optimization from other nature-mimicking methods is that the method is constructive and does not need initial solutions. This helps it more easily depart from local optima and avoid regions of infeasible solutions.

The ant starts building the solution from a random node in the problem graph. It adds a new node using a transition probability. The transition probability $P_{i,j}$, is a product of the heuristic information $\eta_{i,j}$ and the pheromone concentration $t_{i,j}$ linked with the move from node i to the node j, where $i, j = 1, \ldots, n$.

$$P_{i,j} = \frac{t_{i,j}^a \eta_{i,j}^b}{\sum\limits_{k \in Unused} t_{i,k}^a \eta_{i,k}^b}, \tag{1}$$

where $Unused$ is the set of unused nodes of the graph.

The initial pheromone quantity has the same value for all elements of the graph and is a positive value t_0, $0 < t_0 < 1$. After the iteration the ants update the pheromone concentration [4]. A node becomes more attractive if it concentrates more pheromone. a and b are importance of the pheromone and heuristic information respectively.

The main rule for the updating pheromone value is:

$$t_{i,j} \leftarrow r t_{i,j} + \Delta t_{i,j}, \tag{2}$$

where r imitates nature evaporation and pheromone value decrease. $\Delta t_{i,j}$ is a new placed pheromone, which corresponds to the quality of the solution. The quality is estimated by the value of the objective function.

3.2 ACO for Vegetables and Annual Fruit Crops

The application of the method depends on the specific problem. In our application, the ant initially randomly selects one of the crops to seed, then sows it, in the minimum possible amount for that crop, on a randomly selected plot from the seed plots. The next crop to sow and the plot to sow are selected using the transition probability. If there are areas sown with this crop on the respective plot, then an area unit, 1 acre, is added. If there are no areas sown with this crop on the respective plot, the minimum possible area is added. If fewer crops are sown than the minimum number of different crops to be sown, the probability of selecting an unsown crop is doubled.

$$P_{i,j} = \begin{cases} \dfrac{t^a_{i,j}\eta^b_{i,j}}{\sum\limits_{k\in Unused} t^a_{i,k}\eta^b_{i,k}} & \text{more crops or crop } i \text{ is assigned} \\ 2\dfrac{t^a_{i,j}\eta^b_{i,j}}{\sum\limits_{k\in Unused} t^a_{i,k}\eta^b_{i,k}} & \text{less crops, crop } i \text{ isn't assigned} \end{cases} \quad (3)$$

In this bearing we force diversification of sown crops. In opposite, only the most advantageous crop will be sown. In this way, we prevent sowing only the most profitable crop and diversify the sowing. The role of heuristic function is to lead the search in the desired direction. It can contain parameters from the objective function as well as parts of the constraints. In our opinion, the following heuristic information is suitable for the problem being solved:

$$\eta_{i,j} = P0 \times Decr_{i,j} \times d_{i,j} \times P_i + (1 - P0) \times d_{i,j} \times P_i + (S_i - r1_{i,j}) \times N_{ij} \quad (4)$$

where $P0$ is a probability for adverse event, $Decr_{i,j}$ is the yield decrease coefficient in case of adverse event, $d_{i,j}$ is output for crop i from land j, P_i is expected price of crop i, S_i is subsidy for crop i, $r1_{i,j}$ is expenses for crop i from land j, N_{ij} is the sown area of crop i on land j. The proposed heuristic function considers the possibility of an adverse event occurring. It also takes into account the different types of costs of the farmer and the possibilities for subsidies and incentives.

The goals set in solving the problem are reflected by the objective function. In our case, this is maximum profit for the farmer, subject to the available constraints. For this reason, the objective function has a following form:

$$F = \sum_{i=1}^{n}\sum_{j=1}^{M} P0\times Decr_{i,j}\times d_{i,j}\times P_i\times N_{ij} + (1-P0)\times d_{i,j}\times P_i\times N_{ij} + (S_i - r1_{i,j})\times N_{ij} \quad (5)$$

4 Testing and Discussion of the Model

Agricultural crops requiring similar soil types and similar land treatments are usually planted on the same plot. This supports the development of crops and facilitates processing. As an example, we will give that vegetables are not sown at the same plot as cereals.

For our tests, we have chosen zucchini, pumpkins and watermelons. These are crops with similar requirements for soil quality and preliminary treatments. They can be grown close to each other. We fixed the ruling parameters of our ACO model experimentally by several algorithm runs. All tests are made by using same parameters as follows: number of iterations is 100; evaporation parameter $r = 0.5$; initial pheromone values is t_0=0.5; number of used ants is fixed to be 20; $a = b = 1$.

For different types of crops, the minimum seeding land required for cultivation to be effective is different. For cereals, efficient farming is achieved if the minimum sown area is 10 acres. With perennial tree crops, it is not efficient to sow only one acre, usually a much larger area is sown. In the case of vegetables and annual fruit crops, the minimum sown area can be 1 acre. The characteristics of the example used are given in Table 1.

Table 1. Test instances values.

Parameters	Value
parcels of land	$\{100, 200\}$
number of plants	3
minimal area	1
minimal number of vegetables	3

We have prepared a test example containing two plots of 100 and 200 acres respectively. The first plot is more favorable and yields from it are about 5% higher. At the beginning, we consider a case with no probability of an adverse event occurring. In this way, the crops planted are equal.

Table 2. Case without subsidy and adverse events.

parcels	zucchini	pumpkins	watermelons
Parcel1	0	1	99
Parcel2	51	149	0

Analyzing the results, we can note that the cultivation of pumpkins is the most profitable, while the cultivation of zucchini brings less profit, see Table 2. The selling price of zucchini is the highest, but yields are significantly lower than those of pumpkins and watermelons. Pumpkins and watermelons have a lower selling price, but this is compensated by the high yields.

The next case we consider is a 50% chance of drought. Pumpkins have a deep root system and are not greatly affected by drought. Out of the three considered crops, zucchini is most affected by dry weather. For this reason, we have set a 70% reduction in zucchini yields if drought occurs and a 30% reduction in watermelon yields.

The results obtained by the model in this case are shown in Table 3. We observe a decrease in the sowing of courgettes at the expense of an increase in

Table 3. 50% probability for adverse events.

parcels	zucchini	pumpkins	watermelons
Parcel1	0	1	99
Parcel2	1	149	50

the sowing of watermelons. Drought greatly affects zucchini, and the low income from sowing them makes them undesirable for farmers. In the model, we have previously fixed that one crop cannot occupy more than half of all areas. We see that pumpkin seeding fulfills this condition. For this reason, although pumpkins are more profitable, the increase in sowing is with watermelons.

When the probability of drought is small, up to 30%, then it does not affect sowing and it is as it is under normal climatic conditions, see Table 4.

Table 4. Up to 30% probability for adverse events.

parcels	zucchini	pumpkins	watermelons
Parcel1	0	1	99
Parcel2	51	149	0

A probability of only 30% of an adverse event, in this case a drought, is too small to affect the farmers' decision to plant.

Reduced yields due to drought can be compensated by watering. Thus, the yield can become close to the usual one, but the costs of growing the relevant crop increase. Let's have a scenario with a 50% chance of drought. Zucchini are most affected by the unfavorable weather, so they will be additionally watered to obtain the normal yield. The introduction of irrigation to zucchini makes them expensive for growers and from there they become unprofitable and less desirable to plant, see Table 5.

Table 5. 50% probability for adverse events plus irrigation.

parcels	zucchini	pumpkins	watermelons
Parcel1	0	1	99
Parcel2	1	149	50

In this case, irrigation and increased harvest do not contribute to farmers choosing zucchini for sowing. Sowing remains the same as for drought weather.

One of the ways to stimulate the planting of a certain agricultural crop, especially if it is not very attractive to farmers, is subsidy. It should be considered what the amount of the subsidy should be in order to make the agricultural culture attractive. In our example, it is most unattractive for farmers to plant zucchini. Let the zucchini be subsidized and the amount of the subsidy be 150 Euros per acre. Sowing in this case is shown in Table 6.

Table 6. Small subsidy for zucchini.

parcels	zucchini	pumpkins	watermelons
Parcel1	0	1	99
Parcel2	51	149	0

We are observe that sowing is the same as it is without subsidy. Therefore a subsidy of 150 Euros per acre planted with zucchini is too small to make them attractive to farmers. This means that the subsidy must be larger to attract the attention of agricultural producers.

Let's consider a scenario where the courgette subsidy is 250 Euros per acre. In this case, the profitability of the agricultural crop is significantly increased, and this should make it attractive for sowing, Table 7.

Table 7. Higher subsidy for zucchini.

parcels	zucchini	pumpkins	watermelons
Parcel1	0	99	1
Parcel2	150	50	0

We are seeing an increased interest in sowing zucchini. The model suggests half of the area to be planted with zucchini, which is the maximum stake for a single crop. The increased sowing of courgettes comes at the expense of a decrease in sowing of watermelons. Pumpkin seeding remains unchanged as a total. The distribution of the areas with pumpkins by plots is changing. Although the cost of pumpkin seeds is higher than that of watermelons, the selling price is also higher, which offsets the costs. Because of this, the appeal of pumpkins remains unchanged.

In this study, a variety of courgette, pumpkin, and watermelon seeding situations were simulated. The model shows the influence of different parameters on the suitability for sowing of individual crops. The model could have a variety of applications, from helping farmers decide where, what, and how much to grow, taking into account growing costs, land fertility, and weather expectations, to testing the impact of various subsidies, taxes, fees, and regulations.

5 Concluding Remarks

In this work, a seeding model for vegetables and annual fruit crops is proposed. The model is based on ant colony optimization. It takes into account a variety of costs, subsidies and politics, as well as the likelihood of an adverse event occurring. As a test example, the sowing of zucchini, pumpkins and watermelons was used, these are crops with similar soil and processing requirements. The model shows the impact of adverse events such as drought on farmers' willingness to sow the relevant agricultural crop. The impact of subsidies and what their amount should be to increase the attractiveness of a certain agricultural crop has also been investigated. The proposed model can be used to help farmers decide which plot to sow what and in what quantity to have maximum incomes. The model can be used by authorities too, to test a variety of policies to support and restrict given agricultural crops. Seeding of a crop can be encouraged, but within certain limits to prevent oversowing if it becomes too attractive. One can play with the various incentive and constraint mechanisms such as taxes, fees and subsidies. They can be applied both to the area of the sown crop and to the amount of harvest obtained. It also shows the importance and impact of subsidies and how they can be used to incentivize or constrain farmers.

In a future, we will continue developing the model. We will add more details to make it as realistic and useful as possible for both farmers and authorities.

Acknowledgments. The work is supported by the National Scientific Fund of Bulgaria grant DFNI KP-06-N52/5 and by the project COALition №101087022 financed by European Union by the program Horizon Europe/Widening.

References

1. Anagha, K.S., Kuttippurath, J., Sharma, M., Cuesta, J.: A comprehensive assessment of yield loss in rice due to surface ozone pollution in India during 2005–2020: a great concern for food security. Agric. Syst. **215**, 103849 (2024)
2. Bonabeau, E., Dorigo, M., Theraulaz, G.: Swarm Intelligence: From Natural to Artificial Systems. Oxford University Press, New York (1999)
3. Britz, W., et al.: A design for a generic and modular bio-economic farm model. Agric. Syst. **191**(2021), 1–14 (2021)
4. Dorigo, M., Stutzle, T.: Ant Colony Optimization. MIT Press, Cambridge, Massachusetts, USA (2004)
5. Djanibekov, U., Finger, R.: Agricultural risks and farm land consolidation process in transition countries: the case of cotton production in Uzbekistan. Agric. Syst. **164**(2018), 223–235 (2018)
6. Lagerquist, E., et al.: Assessing the effect of intercropped leguminous service crops on main crops and soil processes using APSIM NG. Agric. Syst. **216**, 103884 (2024)
7. Fidanova, S., Roeva, O., Paprzyck,i M., Gepner, P.: InterCriteria analysis of ACO start Startegies. In: Proceedings of the Federated Conference on Computer Science and Information Systems,Gdansk, Poland, pp. 547–550 (2016)
8. Fidanova, S., Luquq, G., Roeva, O., Paprzycki, M., Gepner, P.: Ant Colony Optimization Algorithm for Workforce Planning. In: FedCSIS'2017, IEEE Xplorer, IEEE Catalog Number CFP1585N-ART, pp. 415–419 (2017)
9. Roeva, O., Fidanova, S., Luque, G., Paprzycki, M., Gepner, P.: Hybrid ant colony optimization algorithm for workforce planning. In: FedCSIS'2018, IEEE Xplorer, pp. 233–236 (2018)
10. Fidanova, S., Dimov, I., Angelova, D.: Agricultural system modelling with ant colony optimization. Ann. Comput. Sci. Inf. Syst. **30**, 329–333 (2022)
11. Guo, S., et al.: Adapting crop land-use in line with a changing climate improves productivity, prosperity and reduces greenhouse gas emissions. Agric. Syst. **217**, 103905 (2024)

12. Leip, A., Britz, W., Weiss, F., de Vries, W.: Farm, land, and soil nitrogen budgets for agriculture in Europe calculated with CAPRI. Environmental Pollutionl. **159**, 3243–3253 (2011)
13. Li, N., Li, Y., Yang, Q., Biswas, A., Dong, H.: Simulating climate change impacts on cotton using AquaCrop model in China. Agric. Syst. **216**, 103897 (2024)
14. Lobell, D.B., Burke, M.B., Tebaldi, C., Mastrandrea, M.D., Falcon, W.P., Naylor, R.L:, Prioritizing climate change adaptation needs for food security in 2030. Science **319**, 607–610 (2008)
15. Louhichi, K.: FSSIM, a bio-economic farm model for simulating the response of EU farming systems to agricultural and environmental policies. Agric. Syst. **103**(2010), 585–597 (2010)
16. Mandryk, M., Reidsma, P., van Ittersum, M.K.: Scenarios of long-term farm structural change for application in climate change impact assessment. Landscape Ecol. **27**(2012), 509–527 (2012)
17. Masson-Delmotte, V., et al.: The physical science basis. contribution of working group i to the sixth assessment report of the intergovernmental panel on climate change: summary for policymakers. In: Climate Change 2021, Cambridge University Press, Cambridge, UK (2021)
18. Mucherino, A., Fidanova, S., Ganzha, M.: Introducing the environment in ant colony optimization. Recent Adv. Comput. Optim. Stud. Comput. Intell. **655**, 147–158 (2016)
19. Olesen, J.E., Bindi, M.: Consequences of climate change for European agricultural productivity, land use and policy. Eur. J. Agron. **16**, 239–262 (2002)
20. Ray, D.K., Mueller, N.D., West, P.C., Foley, J.A.: Yield trends are insufficient to double global crop production by 2050. PLoS ONE **8**, 8 (2013)
21. Reidsma, P., Janssen, S., Jansen, J., van Ittersum, M.K.: On the development and use of farm models for policy impact assessment in the European union - a review. Agric. Syst. **159**, 111–125 (2018)
22. Regan, H.M., Colyvan, M., Burgman, M.A.: A taxonomy and treatment of uncertainty for ecology and conservation biology. Ecol. Appl. **12**(2002), 618–628 (2002)
23. Rössert, S., Gosling, E., Gandorfer, M., Knoke, T.: Woodchips or potato chips? How enhancing soil carbon and reducing chemical inputs influence the allocation of cropland, Agricultural Systems **198**(2022), 1–16 (2022)
24. Spiegel, A., Severini, S., Britz, W., Coletta, A.: Step-by-step development of a model simulating returns on farm from investments: the example of hazelnut plantation in Italy: The example of hazelnut plantation in Italy. Bio-based Appl. Econ. **9**, 53–83 (2020)
25. Spiegel, A., Britz, W., Djanibekov, U., Finger, R.: Stochastic-dynamic modelling of farm-level investments under uncertainty. Environ. Model. Sortware **127**, 1–14 (2020)
26. European Union policies, https://ec.europa.eu/info/strategy/priorities-2019-2024/european-green-deal_en?msclkid=86feef9bbafb11ec8c045931e097187b, Accessed 30 Apr 2024

Securing Software Using Genetic Algorithms and Encoding

Alexandru-Gheorghe Reuț(✉)

"Alexandru Ioan Cuza" University, Iasi, Romania
galex.reut@gmail.com

Abstract. Protection of software against reverse engineering that is done by individuals with different purposes becomes more and more complex with time due to evolving technology. No matter how unimportant a piece of software is some kind of protection against intellectual property and data theft must be in place. In the help of protection against reverse engineering, a method of encoding is proposed that encodes the principal section of a Portable Executable(PE) file in such a way that the encoded assembly instruction look like a normal assembly instructions of a program. This method is using a Genetic Algorithm(GA) together with Asm2Vec engine. The Genetic Algorithm has into his population the assembly instructions of the input program and other assembly instructions from other legitimate programs. The decoding is made at the user request based with a key that is uniquely generated for each encoded program. The results show that the encoded program is not obfuscated nor encrypted which may misled the attacker and hide the behavior of the program.

Keywords: Genetic Algorithm · Asm2Vec · security · obfuscation · encoding

1 Introduction

Having a secure software is an important aspect of everyone that offers services trough delivering software on users computers. Many thread actors use reverse engineering to either find vulnerabilities in the software and exploit them or to steal intellectual property or data by knowing how the software is working.

There are many techniques that protect software on all levels(integrity, data, program logic etc.). Some of them that are related to obfuscation will be presented next.

Instructions reordering goes by the idea of permuting instructions in such a way that does not affect the program's initial flow. It's scope is to pass the detection patterns of malicious software. From a human perspective analysis it doesn't add complexity to the program and is somewhat easy to spot the flow of the program [7].

D. Simian and L. F. Stoica (Eds.): MDIS 2024, CCIS 2486, pp. 60–71, 2025.
https://doi.org/10.1007/978-3-031-87386-7_5

Instruction substitution means substitution of some instructions with a sequence of other instructions that do the same thing. The program's semantic is preserved and the size of binary increases. Since there is a limitation for equivalent instructions substitution for a human analysis it will not be hard to spot the logic of the program [7].

Insert dead code consists of inserting blocks of code that is never executed. This blocks of code are introduced to the Control Flow Graph(CFG) of the program by a condition that is always true. Such blocks could be large. This method is modifying the program size [7].

Program encoding refers to the fact that instructions are encoded or encrypted. At runtime each instruction is decoded so that it can be executed and then re-encoded again. The overhead of this technique depends on the algorithm of encoding/encryption but in general is large [7].

For every technique that modifies the size of the binary there are also other changes to the PE file that are need in order for the program to run. This changes may include adjusting the table from .reloc section and the sections shifting to new addresses.

The proposed method may mislead the attacker in some encoded versions into thinking that program he tries to analyze is actually the real one and hide the original behavior of the program. The principal contribution is that the encoded program is not obfuscated nor encrypted, but in fact a key is needed to reveal the original behavior. The key and neither the decoding algorithm aren't embedded into the input program. The decoding happens only on the user's machine. A GA together with Asm2Vec [1] is used to create such encoding. Some of the generated assembly code by the GA may mimic very little of the behavior of the above mentioned techniques.

The structure of the paper is the following. Section 2 will present some related techniques to the one from presented in this article. Section 3 describes in detail the proposed method. Section 4 will show the results of the proposed method. And Sect. 5 gathers some thoughts about the method and future improvements.

2 Related Work

In [8] GAs are used in the field of cryptography to find the best keys for the encryption of images. The encryption algorithms try to convert an image to another image that would be harder to understand. For this the generation of unique keys is needed. GAs evolve the keys such that the image will have much noise therefore will be harder to understand.

In [9] the encryption schemes offer a good protection against static or dynamic analysis and instructions tampering. The encryption schemes consists in self encrypting code. First schema is bulk decryption. The idea is that the entire program is encrypted. When it runs a decryption routine is first executed that decrypt the entire program, similar to a packer. Second scheme is on demand decryption. This schema decrypt and encrypt an entire function when is needed. For this to happen it has a integrity check function that his result is the key. The integrity check is applied on a function or a block of code.

In [10], Javaheri et al. proposes a framework that uses GAs to evolve metamorphic malware to train neural networks so that it will be possible to detect new and unseen malware samples. The GAs here are also used with the scope of increasing the samples number because there aren't many metamorphic malware. In particular GAs are used to evolve sequences of API calls. Then the created sample is tested on secure environment for correctness.

3 Methods

To offer protection of software against reverse engineering an encoding method is developed that it's using GAs to evolve assembly instructions and Asm2Vec [1,2] as GAs fitness function. The proposed method is only for Windows x86 architecture.

From a set of assembly instructions that represent the main section of a PE file it evolves the instructions so that it will may look like the original program instructions. The scope is to mislead the attacker in not realizing that the encoded(encrypted) instructions are in fact fake. The evolved instructions will mimic some kind of logic that may help in this matter based on Asm2Vec [1] similarity value. Also for the instruction like call, jmp, jg, jo, jge etc. after generated, their addresses are adjusted so that they are in the main section(e.g. .text) space address. In the following subsections the method is presented in more details.

3.1 Asm2Vec

The Asm2Vec [1] is an engine that helps analysts to better understand assembly code, either if it's obfuscated or has another method of protection by learning the lexical semantic relationships between blocks of assembly code. This blocks of code can represents functions or specific assembly instructions that may represents an algorithm, logic or just some vital instructions that in the view of the analyst are important. With this in mind the analyst search for similar instructions in a protected(obfuscated) PE [2].

In the proposed method Asm2Vec [1] is used as the fitness function in the GA to get the most similar assembly instructions taking into account the original assembly instructions of the program. The encoding is wanted to be made at instructions level and not on blocks of instructions. Because of this aspect for each input program the Asm2Vec [1] is trained having in it's blocks of code one assembly instruction at a time. This is possible because Asm2Vec [1] doesn't need the correct order of instructions that will be executed. More specifically it's learning the semantic from instruction itself. For example a block of code that contains more assembly instructions given as training, then it will search for permutations of the same pattern that looks almost the same. This behavior can be seen in the Table 1. Sure not all assembly instructions are so semantic descriptive and an understating cannot be accurately made. Thus in some cases

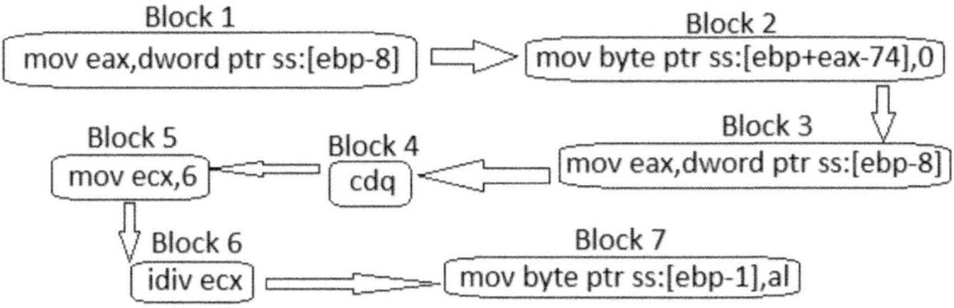

Fig. 1. Asm2Vec blocks of code for training.

like for the instruction *add eax,1* it will find *sub edx,1* to be somewhat similar which in the proposed method is beneficial (Fig. 1).

For these assembly instructions are generated the corresponding similar instructions(encoded). The arrows are just to show the order of execution but they don't matter for Asm2Vec [1].

Table 1. Example of original and evolved assembly instructions.

Original	Encoded
mov eax,dword ptr ss:[ebp-8]	mov eax,dword ptr ss:[ebp-B]
mov byte ptr ss:[ebp+eax-74],0	mov byte ptr ds:[ebx+eax-76],0
mov eax,dword ptr ss:[ebp-8]	mov al,byte ptr ss:[ebp-B]
cdq	cwde
mov ecx,6	mov ecx,5
idiv ecx	idiv al
mov byte ptr ss:[ebp-1],al	mov byte ptr ss:[ebp],al

Table 1 shows how similar generated encoding instructions are. As can be seen they are somewhat very similar in terms of logic.

3.2 Genetic Algorithm

The GA is considered a meta-heuristic that solves problems of search and optimization based on the principle of natural selection where the fittest survives. With the operators such as selection, crossover and mutation inspired from Darwin evolution, it can generate high quality solutions. The scope is to have a diverse population of individuals(solutions) over time(in each iteration).

The population is maintained by how individuals responds to the changes of the environment and how quickly they adapt. The crossover operator maintains a mix of new individuals by combining existing individuals(parents) from the population which brings diversity. The mutation operator also brings a grade of diversity and search to new areas by randomly changing very little in the individual representations(chromosome) [6].

The GAs for each evolved assembly instruction will also generate a key that will be the size of the each original assembly instruction. The key is generated between the original assembly instruction and the evolved assembly instruction found by GAs using the XOR operator. In most of the cases the key is less than the virtual size of the main section. This is because instructions like *int3* or *add byte ptr [eax], al* are not encoded. These instruction like *int3* are only generated by the compiler for debugging purposes.

The decoding of the program, is simply done by XORing the encoded assembly instructions and the uniquely generated key.

Algorithm 1. Encoding algorithm

```
Require: inst, c_r, m_r, iter ← 25
  pop ← 700 instructions                      ▷ randomly chosen(initial population)
  best_encoded_ins ← random(pop)              ▷ a random instruction from the population
  best_score ← fitness(best_encoded_inst)
  while True do
      scores ← fitness(pop)                   ▷ calculate fitness value of every instruction
      best_encoded_inst ← max(best_score, scores)
      if iter = 0 then
          break
      end if
      for instruction in pop do
          best_selected[instruction] ← selection(pop, scores)
      end for
      for instruction in pop do               ▷ create new population
          second_parent ← best_selected[instruction]
          result ← crossover(c_r, inst, second_parent)
          result ← mutation(m_r, result)
          new_population.append(result)
      end for
      pop ← new_population
      iter ← iter − 1
  end while
  return XOR(inst, best_encoded_inst)         ▷ decoding key
```

The parameter *inst* represents the assembly instruction to be encoded, *c_r* crossover rate and *m_r* mutation rate. The *best_selected* variable contains k best solutions, but has the same size as the initial population even if contains duplicates. As can be deduced this method is called for every assembly instruction of the program.

Since the proposed method operates in the context of the Windows PE files some details are needed to be mentioned.

1. The changes that the proposed method does are only on the main section(e.g. .text) of the PE file. Thus the encoded instructions may not be aligned with the rest of the sections in terms of functionality and program logic.
2. The decoding algorithm is not embedded into the encoded program therefore the key is not stored into the program. The whole solution is made with the idea that the user decodes the encoded program with a separate program that receives as input a hash. With the initial program comes the decoding hash and the decoding program. Even if an attacker will analyze the program used for decoding it will not find any usefully information that may be used to succeed in decoding the original program without the key.

Initial Population. For the GA the population contains assembly instructions from original program and assembly instructions parsed from Microsoft sysinternals suite [3] on x86 architecture. The parsing is done using capstone engine [5] that transforms opcode into assembly instructions code. The assembly instructions are organized based on their opcode length. Having in some cases more than enough assembly instructions already parsed it will be taken into the population 700 randomly chosen. The representation of assembly instructions in GA is hold in opcodes. This is because the sequence of opcodes must be interpreted first as assembly instructions and further organize them by opcodes size.

Selection. The selection of solutions from the population is made based on tournament. Basically for each population, random solutions(instructions) are picked. If their fitness value is better than the best so far it will be chosen and at the end of the iteration a possible best solutions is found.

Crossover. The crossover happens randomly in a uniform fashion. Having two parents of the same size, based on the rate of crossover value the opcode will be interchanged for that respective index (Fig. 2). Each time in one of the two parents will always be the assembly instruction given as a parameter from the input program and in the other parent another assembly instruction from the parsed sysinternal suite [3]. At the end of iteration, after the best generated replacement(encoding) is found then a XOR is made between the generated assembly instruction and the instruction(original instruction, first parent).

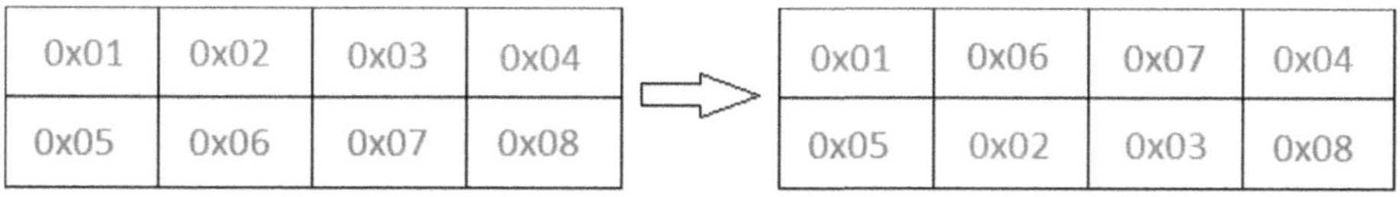

Fig. 2. Example of crossover for instruction of length 4.

After the generation of new children some check is needed. This check consist in verifying that the new generated assembly instructions are valid. Meaning that the generated opcodes represents indeed a valid x86 assembly instruction. For this check is used keystone engine [4].

If the assembly instruction is invalid then the crossover is made again with the same parents.

Mutation. The mutation will happen randomly for only one byte of the input assembly instruction. It consists in subtracting 1 from the byte of a random index of the representation of the assembly instruction.

The same as on crossover, the check for a valid assembly instruction is made again.

This check for valid assembly instructions may seem inefficient and adds more time to the algorithm. The added time on execution is very small(milliseconds), this is because the individuals of the population are different enough from the start and the crossover and mutation rate are higher enough. This ensures that will always find a valid assembly instruction thus each instruction will be encoded.

3.3 Usage of the Proposed Method

On the user's machine that has the proposed method of protection will be the encoded program and a program(named here "Loader") that will do the decoding and also will run the protected PE. The user will receive a hash representing a unique identifier. The decoding key will not be present on the user's machine, only in memory at some point.

For the decoding to happen the users must give the hash as input to the decoding program. The decoding program will send a request with the unique identifier(sha256) to a websites that will send him back the decoding key. Once he has the decoding key it will decode the program, create a new process and run the decoded program (Fig. 3).

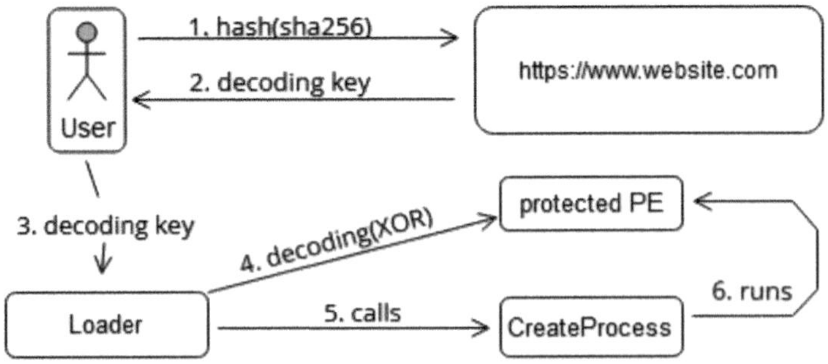

Fig. 3. The process of decoding and execution of protected PE.

4 Results

The below results are made with rate of crossover and mutation equal to 0.7 and with 25 iterations of GA for a single assembly instruction.

4.1 Entropy

First the entropy is calculated for three different programs. The programs are small in terms of size. Also the entropy was calculated using Detect It Easy [14] (Table 2).

Table 2. Results on Entropy.

Program name	Entropy-original	Entropy-encoded v1	Entropy-encoded v2
P1 - 46 asm inst	2.53884 - not packed	2.80170 - not packed	2.74022 - not packed
P2 - 49 asm inst	0.70120 - not packed	0.70324 - not packed	0.71405 - not packed
P3 - 38 asm inst	2.01348 - not packed	2.11671 - not packed	2.17712 - not packed

It can be observed that the entropy of the encoded programs it's increased with a small difference. This can be interpreted as that the proposed method doesn't add nor subtract complexity of the program's original behavior.

4.2 Cyclomatic Complexity(CC)

The Cyclomatic complexity metric is used to calculate the complexity of a program. The CC is calculated based on the CFG of the program [11]. The CC was calculated using Win64dbg [13].

$$M = E - N + 2 * P \tag{1}$$

E means number of edges, N number of nodes and P the number of connected components in the CFG of the program (Table 3).

Table 3. Results on Cyclomatic complexity metric.

Program name	CC-original	CC-encoded v1	CC-encoded v2
P1–46 asm inst	2	3	2
P2–49 asm inst	3	3	3
P3–38 asm inst	4	5	4

It could be said that the results reflect also the results on entropy(very small difference in some cases). No more complexity is added to encoded program which labels it as not obfuscated.

4.3 Safety of the Encoded Programs

To guarantee that the encoded programs are safe to run a test was conducted.

This test consisted of running 10 encoded variants of 2 programs. The tested programs were small having 38 and 49 assembly instructions. When only running and not doing a dynamic analysis of the programs nothing happens to the machine. In the dynamic analysis it showed that some encoded instructions are invalid in the program's context. Meaning that it tries to access addresses from memory or stack that are not mapped into the programs memory space. When this happens it will result in an access violation error and the program will stop. This is because the OS protects other programs memory accesses.

From the point of view of an attacker he may deduce or not that the encoded program must have something else that will help to run it correctly. With this in mind he may try sufficiently enough to modify it such that other assembly instructions will be executed in the hope of understating it. This could happen only because they look almost similar to the original instructions. Also he cannot deduce the original instructions even if the encoded instructions are similar.

4.4 Personal Data Encryption(PDE)

This subsection presents a comparison between PDE [12] and the proposed method.

PDE it's a security solution for encrypting files that are available starting with Windows 22H2(Windows Enterprise and Windows Education). It's using Windows Hello for Business to know encryption/decryption keys for each user credentials. Basically when the user logs off, the files are encrypted. When the user logs in again, the files are decrypted and they can be used again. If another user logs in that files are still inaccessible. Even though is made to offer protection it also has the idea to facilitate the use of their Windows products/features more.

PDE protects against any kind of attack when the files are encrypted. This means that the whole security is happening on the user machine.

The proposed technique protects the program against reverse engineering type of attacks when the program may be on the attacker's machine unless it has the key to decode(key and decoding program) it. But that to happen he must first realize that the program it's actually encoded. Unlike PDE which offers protection only on user's machine.

4.5 Some Examples of Encoded Programs

Second, an example of how encoded instructions looks like versus the instructions of the original program (Fig. 5). And a CFG of the original and encoded program (Fig. 4).

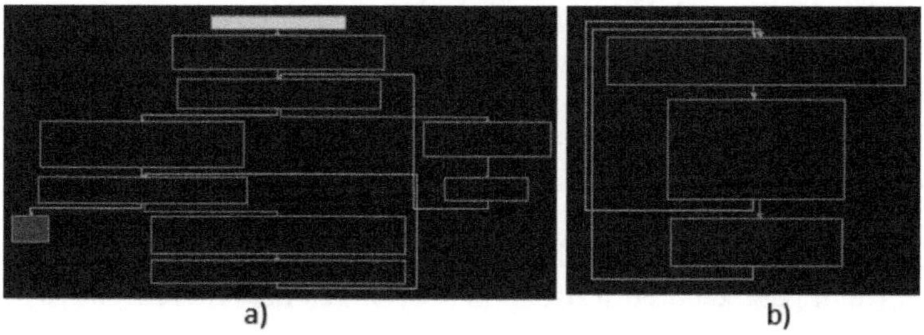

Fig. 4. a) the original program and b) a version of the encoded program.

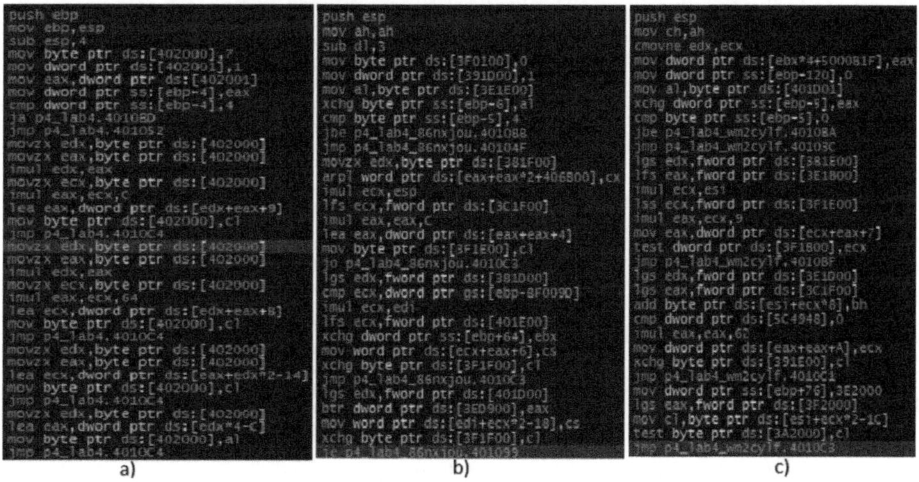

Fig. 5. a) P1 the original program, b) v1 of the encoded program and c) v2 of the encoded program

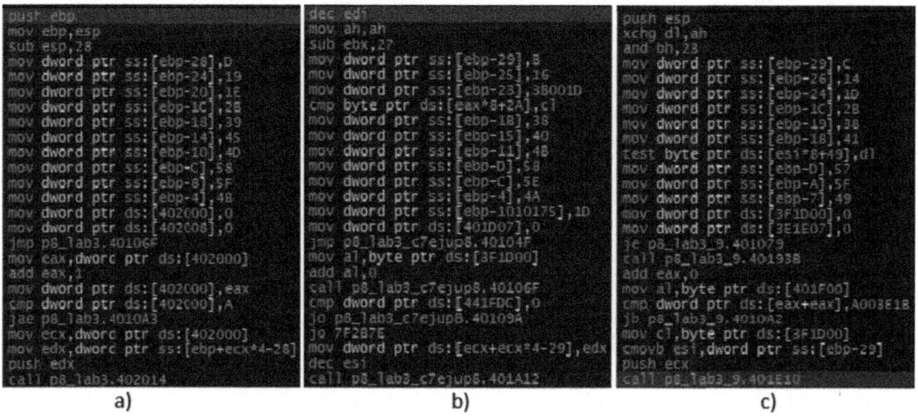

Fig. 6. a) P2 the original program, b) v1 of the encoded program and c) v2 of the encoded program

5 Conclusion and Future Work

The proposed method even if is not so complex it can offer decent security to a piece of software. Even if the attacker will find the key or a small portion of it, it will not know what are the correct instructions compared to the original program. The instructions generated in some portions of the code as can be seen in Fig. 5 and 6, look very alike to the original ones. And also when a large number of them get executed until an exception is encountered, may confuse the attacker even more.

In the future the decoding algorithms could be carefully integrated into the encoded program. The encoded program will receive as input the unique identifier(hash). Send a request with the hash to a websites. The websites will send the key and the decoding will take place. For this to happen a stub(block of code) will always first be executed. To prevent the attacker from understanding it some kind of obfuscation technique must be implemented for it. Also to make the encoded program more indistinguishable, the rest of the sections such as .data and .reloc must be modified in accordance with the generated assembly code from the GA. If the rest of the sections will be modified then the decoding algorithm must also do the decoding for the other sections.

Even if the program will be fully visible into the memory the idea of security goes by the fact that only the user will use the key to decode the software and the decoding program will only be on his computer.

References

1. Asm2Vec GitHub, https://github.com/Lancern/asm2vec
2. Steven, H., Benjamin, C., Philippe, C.: Asm2Vec: boosting static representation robustness for binary clone search against code obfuscation and compiler optimization. IEEE Sympos. Secur. Priv. (2019). https://doi.org/10.1109/SP.2019.00003
3. Sysinternals Microsoft, https://learn.microsoft.com/en--us/sysinternals/downloads/sysinternals--suite
4. Keystone Homepage, https://www.keystone-engine.org/
5. Capstone Homepage, https://www.capstone-engine.org/
6. Eyal, W.: Hands-On Genetic Algorithms with Python. 1st edn. Packt Publishing Ltd, Livery Place 35 Livery Street Birmingham B3 2PB, UK (2020)
7. Sebastian, B., Alexander, P.: A tutorial on software obfuscation. advances in computers 19-24 (2018). https://doi.org/10.1016/bs.adcom.2017.09.004
8. Rajat, J., Vikas, T.: Shivali Dhaka: a study on cryptography using genetic algorithm. Int. J. Comput. Appl. (2015). https://doi.org/10.5120/20860-3559
9. Jan, C., Bart, P.: Towards tamper resistant code encryption: practice and experience. Information Security Practice and Experience (2008), LNSC, vol. 4991, pp. 86–100, Springer, Heidelberg (2008)
10. Danial, J., Pooia, L., Mehdi, H.: A novel method for detecting future generations of targeted and metamorphic malware based on genetic algorithm. IEEE Access (2021). https://doi.org/10.1109/ACCESS.2021.3077295

11. Cyclomatic complexity, https://en.wikipedia.org/wiki/Cyclomatic_complexity
12. Personal Data Encryption, https://learn.microsoft.com/en--us/windows/security/operating--system--security/data--protection/personal--data--encryption/
13. Win64dbg, https://x64dbg.com/
14. Detect it easy, https://www.majorgeeks.com/files/details/detect_it_easy.html

Machine Learning

An Investigative Study Exploring Machine Learning Approaches for Optimizing Deep Brain Stimulation Programming

Esther Ademola[1(✉)], Martin Reich[2], and Magda Gregorova[1]

[1] Center of Artificial Intelligence and Robotics, University of Applied Sciences Wuerzburg-Schweinfurt, Wuerzburg, Germany
esther.ademola@study.thws.de

[2] Department of Neurology, Julius-Maximilians-University Wuerzburg, Wuerzburg, Germany

Abstract. This investigative study explores machine learning models for predicting dystonia improvement scores in Deep Brain Stimulation (DBS). Leveraging data from 85 subjects across seven European DBS centers, we employ various linear and non-linear modeling approaches. In contrast to previous studies utilizing probabilistic mapping, our direct utilization of the actual dataset yields improved results. The random forest model emerges as the most accurate predictor, with a mean deviation of 9.54 ± 6.08%. This implies that for a patient with an improvement score of 78%, the model predicts an improvement between 68% and 87%. This advancement in predictive accuracy holds potential implications for refining DBS programming, ultimately enhancing therapeutic outcomes for individuals with dystonia. In addition, regularization techniques play a pivotal role in determining feature importance thereby contributing to a nuanced understanding of factors influencing DBS therapy outcomes.

Keywords: Deep brain stimulation · dystonia · machine learning · regularization

1 Introduction

Advances in Artificial Intelligence (AI) and Machine Learning (ML), driven by improvements in computational power, data availability, and algorithmic innovation, have revolutionized several sectors, including medicine and healthcare [9]. In medicine, AI & ML have been successfully employed in tasks such as disease diagnosis [19], personalized treatment recommendations, and drug discovery [21], significantly improving the precision and efficiency of healthcare practices.

A particularly interesting application area is neuromodulation, specifically Deep Brain Stimulation (DBS). DBS is an invasive surgical procedure where a neurostimulator is implanted to send electrical impulses to precise areas in the brain, targeting regions involved in neurological function [1]. It is primarily

D. Simian and L. F. Stoica (Eds.): MDIS 2024, CCIS 2486, pp. 75–89, 2025.
https://doi.org/10.1007/978-3-031-87386-7_6

used to treat certain neurological conditions, particularly movement disorders such as Parkinson's disease and dystonia, which are characterized by abnormal involuntary movements, muscle rigidity, and impaired motor control. For example, dystonia involves sustained or repetitive muscle contractions, resulting in twisting movements or abnormal postures. Although these neurological conditions cannot be cured, DBS can significantly improve such motor symptoms by modulating the dysfunctional neural circuits responsible for these abnormalities [7]. The targeted electrical stimulation helps restore more normal neural activity, leading to better motor control and alleviation of symptoms, offering a vital treatment option for patients who do not respond adequately to medication [10].

DBS programming is a sophisticated and iterative process involving the precise adjustment of parameters for the neurostimulator, including stimulation amplitude, frequency, pulse width, and electrode placement. This fine-tuning is essential to optimize therapeutic effects while minimizing adverse effects. Manual programming by clinicians is currently the gold standard for DBS optimization. However, programming DBS for dystonia patients presents significant challenges.

The effects of DBS on motor symptoms typically become apparent only after several weeks or even months, and the lack of dependable biomarkers for therapeutic response complicates the process of acute programming. Consequently, clinicians face the challenging task of configuring pallidal DBS without clear guidance on which target structure will provide optimal symptom relief, depending on the placement of electrodes. Additionally, the delay in observing patient responses, given that symptom improvement can take weeks, leads to significant variability in clinical outcomes [13].

Historically, deep brain stimulation (DBS) programming has relied on clinician expertise and an iterative, trial-and-error process to optimize stimulation parameters. This process can be time-consuming and depends heavily on subjective judgments and extensive clinical experience. Despite its successes, this traditional approach may not capture the full complexity of individual patient responses, which could potentially limit its effectiveness.

Previous research has sought to address these limitations by using probabilistic mapping techniques. By employing probabilistic mapping techniques that utilize electrode positioning and the volume of tissue activation (VTA)-a computational model derived from both DBS programming parameters, including stimulation amplitude, pulse width, and frequency, as well as the localization of the electrode leads [6]—researchers previously conducted voxel-wise mapping of the internal pallidum in 87 patients with dystonia from seven treatment centers [16]. These probabilistic maps enabled the application of a linear regression model to predict the clinical outcomes of DBS based exclusively on individual anatomical features and electrode positioning. This breakthrough led to a successful pilot trial and an ongoing large randomized controlled trial [13].

In our study, we aim to extend this research by bypassing the VTA modeling step and treating DBS programming as a supervised machine learning problem. Rather than relying on traditional linear models, we will employ non-linear regression techniques and ensemble methods, such as random forests and support vector regression, to handle the complexity and dimensionality of real-world DBS datasets. By using patient improvement scores as the target variable, we hypothesize that these machine learning models can capture complex interactions within the data that may not be addressed adequately by linear approaches, potentially providing a more precise, individualized framework for DBS programming.

In this study, we aim to answer the following research question: How can patient-specific data, combined with machine learning models and advanced algorithms, be leveraged to optimize Deep Brain Stimulation (DBS) programming, improving clinical outcomes and reducing the burden on clinicians? To enhance the underlying predictive models, we opted to work with the raw dataset, that is numerical and categorical programming parameters, instead of VTAs which are derived from the programming parameters. The decision is motivated by the potential information loss during the conversion of programming parameters to VTAs as they are just a simulated estimate.

The models chosen for this study include Lasso, ridge regression, support vector regression (SVR), random forest, and XGBoost. These models were selected based on their combined ability to address the challenges specific to DBS programming outcome prediction, especially given the small sample size and high-dimensional nature of the data. In medical research, particularly with complex interventions like DBS, datasets are often limited, which complicates predictive modeling. To make optimal use of our data, we employed a leave-one-out cross-validation (LOOCV) approach, where each model was trained on all samples except one, which was used for testing. This cross-validation technique minimizes overfitting risk, thereby enhancing model generalizability while maximizing the use of our available data.

Each model was selected to balance predictive power with interpretability, addressing both non-linearity and the need to identify influential predictors of DBS outcomes. Linear models, such as Lasso and ridge regression, offer interpretability and aid in feature selection and regularization, which is valuable for highlighting key predictors without excessively simplifying the model. On the other hand, non-linear models such as SVR, Random Forest, and XGBoost are included to manage the complex interactions and subtle non-linear patterns that are typical in DBS response data, where individual patient variability plays a significant role. Ensemble methods like Random Forest and XGBoost, in particular, were chosen for their ability to enhance predictive accuracy and robustness by capturing intricate relationships among stimulation parameters, anatomical features, and clinical outcomes. Together, this suite of models allows for a comprehensive exploration of DBS outcome predictors, leveraging both linear and non-linear techniques to improve prediction precision and offer insights into individual patient responses to DBS.

To support reproducibility and transparency, we provide the code used in this study in an open-access GitHub repository[1]. While the patient-specific data cannot be shared due to its sensitivity, the repository includes the methodology, models, and hyperparameters used.

---

[1] https://github.com/IfeAdemola/ml-driven-dbs-programming.

2 Literature Review

DBS has revolutionized the treatment of neurological disorders, particularly Parkinsonian diseases, by providing an effective therapeutic approach. Traditionally, DBS programming has been a labor-intensive, clinician-driven process that often requires multiple adjustment sessions based on clinical trial and error. However, advancements in machine learning (ML) have introduced novel methods for DBS optimization, reducing the dependency on manual tuning and potentially increasing the precision and personalization of treatment.

The application of ML to DBS programming has grown, with early studies demonstrating its potential to streamline and improve the clinical decision-making process. Khojandi et al. conducted a notable retrospective study on 20 patients who underwent Subthalamic Nucleus (STN)-DBS, analyzing preoperative patient data and disease characteristics to predict optimal stimulation frequency. Using a random forest model trained on preoperative metrics and clinical scores, they achieved 95% accuracy in predicting optimal DBS settings for Parkinsonian symptoms as measured by the Unified Parkinson's Disease Rating Scale (UPDRS) III [11,14]. This study illustrates the utility of ML in modeling DBS settings based on patient-specific data, indicating that data-driven predictions could expedite the time-intensive fine-tuning process for improved patient outcomes. Similarly, Shamir et al. explored ML applications for DBS and medication management by training several models, including support vector machines (SVM), Naive Bayes (NB), and random forests, on data from 89 post-DBS follow-up visits. Their objective was to predict post-DBS improvements in motor symptoms and to assist in medication titration, achieving 86% accuracy in motor outcome predictions one year post-surgery. The study found that SVM performed best in predicting tremor and speech outcomes, while random forests were more accurate for axial akinetic symptoms [18]. Shamir et al.'s clinical decision support system underscores the promise of ML not only for optimizing DBS parameter settings but also for managing medication in tandem, enhancing overall patient care by modeling both device and pharmacologic interventions.

Expanding on these initial ML applications, Reich et al. investigated DBS for dystonia patients through probabilistic mapping within the pallidal region, aiming to enhance therapeutic efficacy by optimizing stimulation volume. Utilizing a simulation software to visualize DBS lead positions and normalize the Volume of Tissue Activation (VTA) into a standardized reference space, they evaluated electrode contacts based on clinical efficacy, employing statistical tests (two-sample t-tests) to establish relationships between electrode placement and patient outcomes [16]. By generating probabilistic maps of antidystonic effects, they highlighted the impact of electrode localization on motor outcomes, with predictive modeling via linear regression and leave-one-out cross-validation yielding an average prediction error of $16.9 \pm 11.6\%$ in motor symptom improvement. This approach highlighted the importance of precise electrode targeting, as their probabilistic maps emphasized that location specificity within the internal pallidum was crucial for maximizing therapeutic effects.

Despite the promising outcomes of these studies, most rely heavily on VTAs or simulated approximations of DBS effects rather than direct data from actual programming parameters. Reich et al.'s reliance on VTA models introduces a level of abstraction that, while valuable, may not fully capture the nuances of real-world DBS programming. VTAs, as computational models, involve inherent estimation and the potential loss of detailed information critical to patient-specific optimization.

To address these limitations, our approach diverges by directly utilizing actual DBS programming parameters and lead localization data. This shift from VTA-based modeling is motivated by the recognition that simulation-based estimations may omit critical details necessary for highly accurate and individualized DBS programming. By modeling from the raw programming data, our study aims to retain the full spectrum of available information, providing a more detailed and accurate basis for predictive modeling. This approach is anticipated to enhance the accuracy and granularity of ML-driven DBS programming models, providing a comprehensive understanding of factors influencing DBS outcomes in dystonia.

3 Materials and Methods

3.1 Data Description

This retrospective study encompasses datasets from 85 subjects diagnosed with dystonia who underwent chronic treatment with bilateral pallidal Deep Brain Stimulation (DBS) across seven European DBS centers [16].

Each patient was implanted with quadripolar macroelectrodes, specifically models 3389 or 3387 from Medtronic Inc., targeting the postero-ventral Globus Pallidus interna (GPi) in the brain. In order to maintain a standardized assessment, video recordings of all patients before and after surgery were collected and retrospectively evaluated. To objectively assess the severity of dystonia and the effectiveness of DBS therapy, established clinical rating scales were employed. For individuals with cervical dystonia, the Toronto Western Spasmodic Torticollis Rating Scale (TWSTRS) [5] was used, while the Burke-Fahn-Marsden Dystonia Rating Scale (BFMDRS) [12] was applied for those with generalized dystonia (Fig. 1). For comparative analysis, the results were normalized by determining the percentage change in both TWSTRS and BFMDRS scores before and after DBS therapy. This approach provided a uniform metric for evaluating treatment effectiveness across the study population.

The dataset includes the patients' age, gender, brain hemisphere, stimulation frequency, amplitude and pulse width, the position of the lead electrode based on the coordinates of the tip and the top of the electrode. It also includes categorical information on the grounded and activated contact points. Also available is the lead electrode type and the transformation matrix that converts the coordinate system from the electrode space to the brain MNI space [3].

3.2 Data Preprocessing

In preparation for predictive modeling, a preprocessing workflow was developed to ensure that the dataset was both robust and standardized across multiple dimensions. Initial feature selection was performed with close alignment to prior studies, particularly that of Reich et al. [16], to retain comparability in the analysis. Given that age and gender were not included as a predictor in earlier research, this feature was excluded from the dataset. Instead, we focused on clinical and technical features of primary relevance, including stimulation parameters (frequency, amplitude, and pulse width), electrode position data, and categorical attributes related to the electrode type (Medtronic models 3389 or 3387) and contact activation status.

To further prepare the data, transformation of electrode coordinates was necessary to map each electrode's local coordinate system to a common anatomical space, providing spatial consistency across patients. Each subject's data included a 4×4 transformation matrix specific to their electrode setup, which was used to align coordinates with either the MNI brain space [3]. The transformation involved parsing the matrix from its string representation and verifying dimensional compatibility, ensuring the transformation accurately mapped the electrode positions to a shared anatomical framework. Through this process, we generated standardized spatial coordinates x, y, z for each electrode tip and active contact, enabling comparability in electrode positioning across the patient cohort.

Building upon this transformation, we then calculated the position of the active contact midpoint for each electrode, offering a representative spatial reference for therapeutic intervention within the target brain region. The Medtronic models 3389 and 3387 possess fixed inter-contact distances (2 mm and 3 mm, respectively), which were considered when determining the midpoint of activated contacts. In cases where two consecutive contacts were activated, the midpoint was calculated, and the resulting coordinates were transformed into the anatomical space using the aforementioned transformation matrix. This approach established a consistent and biologically relevant coordinate for therapeutic localization, facilitating a nuanced analysis of active contact placement across various electrode types.

With spatial alignment complete, redundant or aggregate features were systematically removed to streamline the dataset and minimize feature redundancy. Columns containing aggregate and intermediary data, were identified and excluded unless specifically designated for analysis. This step reduced potential multicollinearity in subsequent modeling stages, ensuring that only essential, non-redundant features were retained for analysis.

Normalization of numerical features followed, addressing potential scale discrepancies that could impact model performance. To standardize continuous variables, particularly the DBS programming parameters (amplitude, frequency, and pulse width), we applied Z-score normalization, transforming each feature to have a mean of zero and a standard deviation of one. This normalization step mitigated the influence of varying scales across features, ensuring that each variable contributed comparably to the model without introducing bias due to scale differences.

The final stage of preprocessing involved encoding categorical variables to render them compatible with modeling requirements. The dystonia type was converted into numerical labels through ordinal encoding, which preserved essential distinctions among categories while enabling their seamless integration into the dataset. This transformation ensured that categorical features were appropriately recognized by the model without compromising the interpretive power of the data.

To facilitate reproducibility and streamline subsequent analyses, a column transformation pipeline was implemented to standardize feature processing for both training and inference datasets. This pipeline incorporated both normalization of numerical features and encoding of categorical variables, ensuring consistent preprocessing at each stage of analysis. The transformation object was retained for future inference, establishing a reliable and reproducible approach to data handling throughout the study (Table 1).

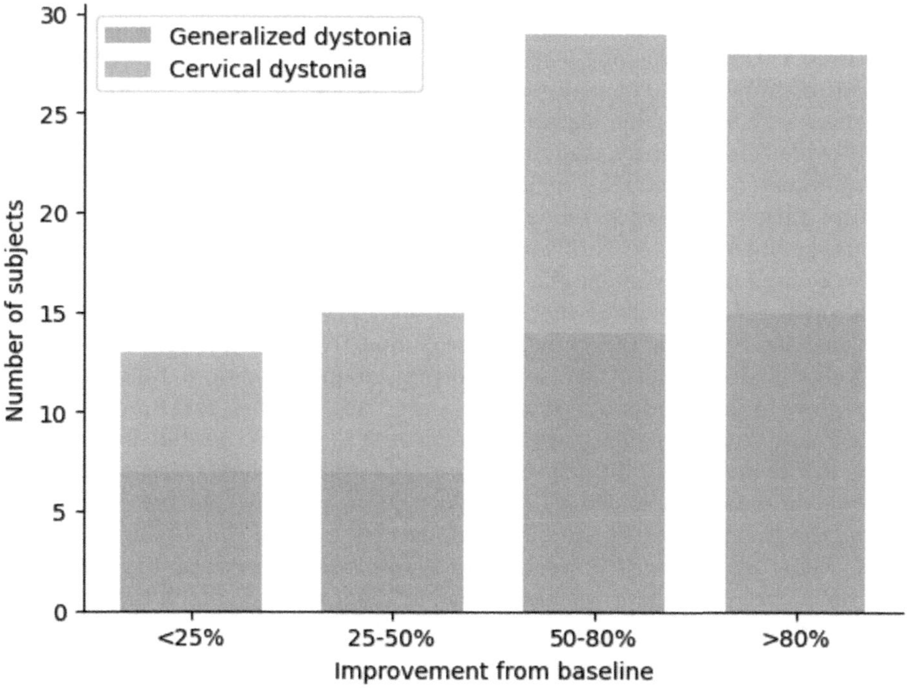

Fig. 1. Improvement observed in TWSTRS for cervical or BFMDRS for generalized dystonia when compared to the initial assessment. Patient's brain hemispheres score were classified into one of four response categories: non-responder (<25%), average responder (25–50%), good responder (50–80%), and super-responder (>80%).

Table 1. Statistical description of dataset.

Characteristics	Cervical dystonia	Generalized dystonia
Number of patients	42	43
Improvement score, %	59.4 (29.3)	62.4 (31.5)
Stimulation parameters		
Amplitude, mA	3.4 (1.1)	3.4 (1.1)
Pulsewidth, μs	109.7 (47.0)	100 (34.9)
Frequency, Hz	153.4 (27.8)	146.9 (29.2)

Data are presented as mean (standard deviation).

3.3 Methodology

This study replicates and builds upon previous research to analyze DBS outcomes in dystonia patients, leveraging Python-based machine learning tools to re-evaluate established findings. The original implementation relied on a MATLAB-based tool called Arena [17], which presented compatibility challenges when replicated directly in Python. Differences in coordinate systems and limitations in MATLAB's machine learning toolset required careful adaptation, leading us to re-implement key components using Python libraries. Python's ecosystem offers diverse machine learning packages, such as scikit-learn and xgboost, that provide comprehensive support for data preprocessing, model selection, and hyperparameter tuning. This flexibility allowed us to systematically explore and optimize various regression models suited to our dataset's characteristics and the study's objectives.

To predict DBS treatment efficacy and identify patterns in patient outcomes, we selected a diverse set of machine learning models: Lasso (Least Absolute Shrinkage and Selection Operator), ridge, support vector regression (SVR), random forest, and XGBoost. Each model brings unique advantages tailored to specific aspects of the dataset and research goals. Lasso [20] and ridge regression [8], as linear models with regularization, were chosen to address multicollinearity and ensure model interpretability, both of which are critical when working with high-dimensional data. Lasso, using L1 regularization, performs feature selection by driving the coefficients of less influential features to zero, simplifying the model and focusing on the most impactful predictors. Ridge, with L2 regularization, constrains coefficient sizes without reducing them to zero, which helps manage multicollinearity while preserving all variables in the model. These regularization techniques are effective in reducing overfitting and are particularly useful in medical datasets, where variable relationships can be complex but interpretable models are highly valued for clinical insights.

To capture potential non-linear interactions between variables, we included SVR [15] with a radial basis function (RBF) kernel. SVR is capable of modeling complex, non-linear relationships that may exist between DBS settings and patient outcomes. The RBF kernel enables SVR to map input features into higher-dimensional spaces, allowing for the identification of subtle interactions that linear models might miss.

Recognizing the value of ensemble learning, we also integrated Random Forest and XGBoost models, both of which have shown success in high-dimensional medical data for their robustness and ability to reduce overfitting through aggregation techniques. Random forest [2] is an ensemble of decision trees that averages multiple tree outputs, reducing the variance associated with individual decision trees and improving generalization. XGBoost [4], an advanced gradient boosting method, was included for its efficiency in handling complex datasets with many features and its ability to minimize prediction errors iteratively. Ensemble methods are valuable in this context because they combine the strengths of multiple models, capturing a wide array of data patterns, while minimizing the likelihood of overfitting-a concern given the study's limited sample size. Ensemble approaches also contribute to model interpretability, as their outputs help provide a comprehensive perspective on how different model architectures respond to the data.

Hyperparameter tuning for each model was conducted using GridSearchCV, a systematic approach that iteratively explores combinations of hyperparameters to identify the best-performing configurations. Given the limited size of our dataset, Leave-One-Out Cross-Validation (LOOCV) was chosen as the evaluation technique to maximize the use of available data while ensuring robust model assessment. LOOCV involves using each individual data point as a validation set, with the remaining data points as the training set, cycling through each point in the dataset. Although computationally intensive, LOOCV is advantageous for small datasets as it reduces estimation bias and provides an almost unbiased evaluation of model performance. This method allowed us to tune hyperparameters while avoiding data wastage, ultimately leading to more reliable performance estimates despite the sample size constraints.

To assess model performance and stability, we evaluated each model using root mean square error (RMSE) as the primary metric, which prioritizes minimizing the variance between predicted and actual values. RMSE is sensitive to larger errors, which is critical in medical applications where accuracy is paramount. Additionally, we analyzed feature importance using the coefficients from Lasso and Ridge regressions. By examining the relative contributions of each feature in these regularized models, we identified which variables most strongly associated with positive DBS outcomes. This interpretative step was essential for translating the findings into insights that could aid clinicians in personalizing DBS settings.

The methodology applied in this study—combining linear (Lasso, Ridge), non-linear (SVR), and ensemble (Random Forest, XGBoost) models provided a well-rounded exploration of predictor variables, allowing for a thorough examination of both simple and complex relationships within the data. By carefully tuning and validating each model, we aimed to maximize predictive accuracy, thereby enhancing the potential for precise DBS programming recommendations. The following section presents the results of these analyses, comparing model performance and discussing their implications for future DBS treatment planning.

4 Results and Discussion

The evaluation of machine learning models for predicting dystonia improvement scores in DBS therapy reveals distinct patterns in model performance. Table 2 shows that the random forest model achieved the highest accuracy, with a mean deviation of 9.54% ± 6.08% from actual improvement scores. This represents a substantial improvement over the baseline model by Reich et al., which had a mean deviation of 16.9% ± 11.6%. Notably, random forest's performance was considerably better than that of the linear models (Lasso and ridge regression), SVR, and XGBoost, each of which demonstrated higher mean deviations from actual scores.

The superior accuracy of random forest is likely due to its ability to capture complex, non-linear interactions within the dataset without assuming a strictly linear relationship between features and outcomes. Random forest's ensemble nature allows it to combine multiple decision trees, reducing overfitting and improving generalization—an advantage in small datasets like ours, where overfitting is a concern.

In contrast, the other models encountered challenges that limited their effectiveness. Both Lasso and ridge regression, as linear models, assume straight-line relationships between predictors and outcomes, making them less capable of capturing the intricate, non-linear interactions likely present in DBS therapy outcomes. Despite their regularization techniques, these models struggled to balance interpretability with predictive power, especially in a dataset with complex feature relationships.

Support vector regression with an RBF kernel, while non-linear, requires careful hyperparameter tuning and is sensitive to outliers. Given our small dataset, tuning parameters effectively was challenging, which may have prevented SVR from capturing subtle patterns in the data. Additionally, SVR's sensitivity to outliers could have skewed predictions, especially with the inherent variability in patient responses to DBS therapy.

XGBoost, a gradient boosting model, showed intermediate performance but was likely affected by its tendency to overfit on small datasets. Although powerful in larger datasets, XGBoost's sequential approach of refining predictions on errors can lead to overfitting when data is limited, especially if the model "learns" noise rather than true patterns. Fine-tuning regularization parameters on our dataset may not have fully mitigated this issue, resulting in relatively higher prediction deviations compared to random forest.

Moving beyond performance metrics, the analysis of feature importance using the aforementioned regularization methods provides valuable insights into the factors influencing DBS therapy outcomes. With the L1 and L2, the y-coordinate of the electrode tip, pulse width, the z-coordinate of the active contact well as the dystonia type emerged as pivotal features, exhibiting significant coefficients

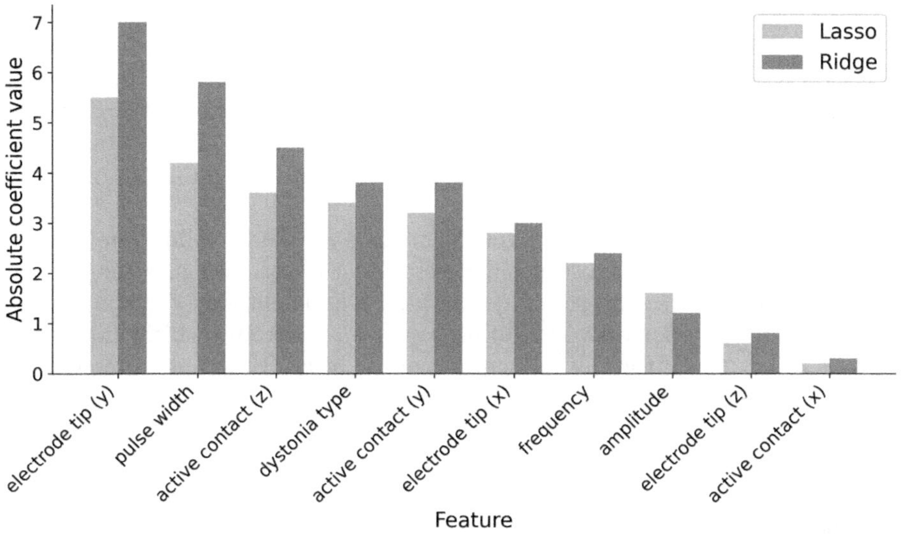

Fig. 2. The absolute coefficient value reveals the importance of the features in the dataset, with a higher coefficient relating to a stronger correlation of the feature and the target variable, dystonia improvement score.

(Fig. 2). The type of dystonia, a patient-specific characteristic, further underscores the need for personalized approaches in DBS therapy. Unsurprisingly, the frequency and amplitude of the electric impulse appeared to have less influence on predicting dystonia improvement scores. This was also highlighted in the study by Reich et al. where the linear regression analysis showed no significant correlation between the stimulation frequency and the patient's improvement [16].

Table 2. Mean difference and standard deviation of predictions from actual improvement scores.

Model	Mean Difference $\pm$ Standard Deviation(%)
Baseline (Reich et al.)	16.9 $\pm$ 11.6
Lasso	24.57 $\pm$ 14.92
Ridge regression	24.16 $\pm$ 15.17
Support vector	23.6 $\pm$ 16.96
Random forest	**9.54** $\pm$ **6.08**
XGBoost	19.50 $\pm$ 12.48

While our study presents improvements in using machine learning models to predict DBS therapy outcomes, several limitations could impact the validity and generalizability of the findings. First, the sample size of 85 subjects may not capture the full diversity of dystonia patients, limiting the applicability of the results to broader populations with varying clinical characteristics. Additionally, the exclusion of patients with incomplete data introduces potential bias. Another concern is the sensitivity of the data, which restricts its availability for replication, posing challenges for external validation despite the public availability of the code.

Although LOOCV provided a robust method for model validation, it is computationally intensive for a larger dataset, and the variance in error estimates may still affect the reliability of the results. The exclusion of certain unmeasured variables, such as comorbidities, could also introduce bias, limiting the model's ability to capture all influential factors in DBS therapy outcomes. Furthermore, temporal aspects, such as disease progression and long-term effects of DBS, were not incorporated into the model, potentially restricting its predictive power. Future work should address these limitations by including larger, more diverse datasets, exploring alternative cross-validation techniques, and integrating temporal and demographic factors to enhance the precision and robustness of DBS programming models.

To summarize this section, in this study the random forest model stands out as a robust predictor of DBS therapy outcomes, showcasing its ability to navigate the non-linear relationships within the dataset. The identified key features, including localization of the electrode, pulse width, and dystonia type, provide nuanced insights into the intricate dynamics of DBS treatment. Acknowledging the limitations, such as the computational demands of LOOCV, paves the way for future research to refine methodologies and explore more efficient validation techniques. These findings collectively contribute to the ongoing discourse on enhancing the precision and efficiency of DBS therapy for dystonia patients.

5 Conclusion

In conclusion, our investigation into machine learning models for predicting dystonia improvement scores in DBS therapy unveils compelling insights. Leveraging a dataset of bilateral therapy from 85 subjects and employing both linear and non-linear models, our study achieves a notable breakthrough with the random forest model, exhibiting a predicted mean deviation of $9.54 \pm 6.08\%$.

The integration of regularization techniques, including Lasso and Ridge Regression, plays a pivotal role in identifying key features influencing DBS therapy outcomes. Specifically, the localization of the electrode in the y direction, pulse width, and dystonia type emerge as critical determinants, offering a nuanced understanding of the intricate factors shaping treatment responses. Additionally, we found that stimulation frequency and amplitude have an insignificant effect on patient improvement.

Our results propose potential refinements in DBS programming, emphasizing a tailored approach based on the identified key features. This insight holds promise for personalized and optimized therapeutic interventions, potentially revolutionizing clinical practices by minimizing side effects and enhancing treatment efficacy.

Looking ahead, we aim to incorporate the age of disease onset and the age at surgery into our analysis. This inclusion is anticipated to provide valuable insights, potentially improving the accuracy and effectiveness of our modeling efforts. Furthermore, the application of other AI tools could significantly enhance the evaluation process. Currently, the symptom severity ratings are conducted by clinicians, which can be error-prone due to their reliance on visual assessments that may vary based on the clinician's experience and the specific circumstances of the evaluation. Such variability introduces bias, influenced not only by the clinician but also by the day and context of the assessment.

The introduction of AI techniques, such as human recognition, task recognition, and activity tracking, could provide objective measurements of kinematic metrics, including speed of movement and frequency, allowing for more accurate symptom severity scores. This shift would mitigate the inherent biases associated with clinician-based evaluations.

Moreover, we plan to integrate longitudinal data from follow-up visits to understand the impact of programming parameters over time. Investigating the combined effects of DBS and medication on symptoms will also be crucial, especially when considering different configurations (DBS on/off, medication on/off) and their interactions. This multifaceted approach holds the potential not only to enhance DBS programming but also to improve the overall management of neurological disorders through more tailored therapeutic interventions.

In essence, our study not only advances the understanding of DBS therapy but also envisions a future where clinical practices are informed by nuanced insights into key features. By decoding the intricate relationships between these features and therapy outcomes, we anticipate a paradigm shift towards more precise, tailored, and effective DBS programming for individuals grappling with dystonia.

Acknowledgments. I would like to express my sincere gratitude to Prof. Dr. Martin Reich and his dedicated team at VisualDBS Lab, Department of Neurology, University of Würzburg, for their invaluable support in providing the data for this research. Their expertise and patience in guiding me through the domain knowledge have been instrumental in shaping the foundation of this study.

I am also deeply thankful to Prof. Dr. Magda Gregorová for her exemplary supervision throughout the course of this study. Her insightful feedback, encouragement, and mentorship have played a pivotal role in shaping the progress of this study.

Disclosure of Interests. The authors have no competing interests to declare that are relevant to the content of this article.

References

1. Benabid, A.L.: Deep brain stimulation for Parkinson's disease. Curr. Opin. Neurobiol. **13**(6), 696–706 (2003). https://doi.org/10.1016/j.conb.2003.11.001
2. Breiman, L.: Random forests. Mach. Learn. **45**, 5–32 (2001). https://doi.org/10.1023/A:1010933404324
3. Brett, M., Johnsrude, I.S., Owen, A.M.: The problem of functional localization in the human brain. Nat. Rev. Neurosci. **3**, 243–249 (2002). https://doi.org/10.1038/nrn756
4. Chen, T., Guestrin, C.: Xgboost: a scalable tree boosting system. In: Proceedings of the 22nd ACM SIGKDD International Conference on Knowledge Discovery and Data Mining, vol. 11, pp. 785–794. ACM (2016). https://doi.org/10.1145/2939672.2939785
5. Consky, E.: The toronto western spasmodic torticollis rating scale (TWSTRS): assessment of validity and inter-rater reliability. Neurology **40**(Suppl 1), 445 (1990)
6. Frankemolle, A.M., et al.: Reversing cognitive-motor impairments in Parkinson's disease patients using a computational modelling approach to deep brain stimulation programming. Brain **133**, 746–761 (2010)
7. Hariz, M., Blomstedt, P.: Deep brain stimulation for Parkinson's disease. J. Intern. Med. **292**(5), 764–778 (2022). https://doi.org/10.1111/joim.13541
8. Hoerl, A.E., Kennard, R.W.: Ridge regression: biased estimation for nonorthogonal problems. Technometrics **42**(1), 80–86 (2000). https://doi.org/10.2307/1271436
9. Jiang, F., et al.: Artificial intelligence in healthcare: past, present and future. Stroke Vasc. Neurol. **2**(4), 230–243 (2017). https://doi.org/10.1136/svn-2017-000101
10. Jinnah, H.A., Factor, S.A.: Diagnosis and treatment of dystonia. Neurol. Clin. **33**(1), 77–100 (2015). https://doi.org/10.1016/j.ncl.2014.09.002
11. Khojandi, A., Shylo, O., Mannini, L., Kopell, B.H., Ramdhani, R.A.: Stratifying Parkinson's patients with STN-DBS into high-frequency or 60 Hz-frequency modulation using a computational model. Neuromodulation: Technol. Neural Interface **20**(5), 450–455 (2017). https://doi.org/10.1111/ner.12607, https://www.sciencedirect.com/science/article/pii/S109471592103316X
12. Kuiper, M.J., et al.: The burke-fahn-marsden dystonia rating scale is age-dependent in healthy children. Mov. Disord. Clin. Pract. **3**(6), 580–586 (2016). https://doi.org/10.1002/mdc3.12339
13. Lange, F., et al.: Dips (dystonia image-based programming of stimulation): a prospective, randomized, double-blind crossover trial. Neurol. Res. Pract. **3** (2021). https://doi.org/10.1186/s42466-021-00165-6
14. M.D.S.T.F.: On rating scales for Parkinson's disease.: the unified Parkinson's disease rating scale (UPDRS): status and recommendations. Mov. Disord. **18**, 738–750 (2003). https://doi.org/10.1002/mds.10473
15. Picard, R.R., Cook, R.D.: Cross-validation of regression models. J. Am. Stat. Assoc. **79**(387), 575–583 (1984). https://doi.org/10.2307/2288403
16. Reich, M., et al.: Probabilistic mapping of the antidystonic effect of pallidal neurostimulation: a multicentre imaging study. Brain **142**, 1386–1398 (2019). https://doi.org/10.1093/brain/awz046
17. Roothans, J.: Arenatoolbox (2024). https://github.com/JonasRoothans/ArenaToolbox, gitHub repository. Accessed 13 Nov 2024
18. Shamir, R.R., Dolber, T., Noecker, A.M., Walter, B.L., McIntyre, C.C.: Machine learning approach to optimizing combined stimulation and medication therapies for Parkinson's disease. Brain Stimulation 8(6), 1025–1032 (2015). https://doi.

org/10.1016/j.brs.2015.06.003, https://www.sciencedirect.com/science/article/pii/S1935861X15010049

19. Singh, P., Singh, N., Singh, K.K., Singh, A.: Chapter 5 - diagnosing of disease using machine learning. In: Singh, K.K., Elhoseny, M., Singh, A., Elngar, A.A. (eds.) Machine Learning and the Internet of Medical Things in Healthcare, pp. 89–111. Academic Press (2021). https://doi.org/10.1016/B978-0-12-821229-5.00003-3, https://www.sciencedirect.com/science/article/pii/B9780128212295000033
20. Tibshirani, R.: Regression shrinkage and selection via the lasso. J. Roy. Stat. Soc. Ser. B (Methodol.) **58**(1), 267–288 (1996). http://www.jstor.org/stable/2346178
21. Zhavoronkov, A., Vanhaelen, Q., Oprea, T.I.: Will artificial intelligence for drug discovery impact clinical pharmacology? Clin. Pharmacol. Ther. **107**(4), 780–785 (2020). https://doi.org/10.1002/cpt.1795

Pretrained Language Models for Automatic Source Code Generation: Evaluation and Application

Adrian Bender and Santiago Nicolet(✉)

Universidad del Salvador, C1023 Buenos Aires, Argentina
{bender.adrian,santiago.nicolet}@usal.edu.ar

Abstract. This study explores the capabilities of the ChatGPT artificial intelligence language model, with the GPT-3 architecture developed by OpenAI, in the task of generating JavaScript source code from Spanish instructions. Transformer language models, as exemplars of deep learning, are effective at learning contextual representations of words and phrases. This allows the model to understand not only individual programming terms but also how they combine into larger structures such as loops and functions. Through a set of unique programming feature requests from a selected set of cases prepared for this work, we examine the model's ability to transcribe these high-level specifications into executable and functional source code. The results of the model were evaluated using a compiler, seeking an objective evaluation of the functionality of the generated code in predefined unit test cases. Overall, the model achieves 100% compilable code and 90% successful individual problem resolution. This work explores the intersection of AI and programming, paving the way for effective automation of code development from sentences in Spanish. While our analysis is centered on Spanish, the insights gained could be relevant for other non-English languages as well. It is expected that this work will contribute to the growing body of literature focusing on code generation and natural language understanding in AI language models.

Keywords: Artificial Intelligence · Transformer Models · Source Code Generation

1 Introduction

1.1 Status of AI Models

Advances in AI in recent years have been assisting developers throughout the software development lifecycle. It's not that programmers are being replaced by robots; rather, AI-driven tools are making them more productive and effective [1].

Our research focuses on determining the capabilities of text generation models based on pretrained language models for generating correct source code. Additionally, we examine the effectiveness of translating instructions from Spanish into programming instructions, providing a unique perspective on the natural language understanding capabilities of Transformer-based models [2].

D. Simian and L. F. Stoica (Eds.): MDIS 2024, CCIS 2486, pp. 90–104, 2025.
https://doi.org/10.1007/978-3-031-87386-7_7

This work promises to unlock new avenues of exploration at the intersection of artificial intelligence and programming, paving the way for more effective automation of software development and greater accessibility for non-English languages. Consequently, this study is expected to provide a valuable contribution to the growing body of literature that focuses on automatic code generation and natural language understanding in AI language models.

In our initial investigation of tools for automatic source code generation through pretrained language models [3], it was revealed that these tools could generate simple programs from text strings referred to as prompts in English or documentation-style headers. The capabilities of different tools and their results were particularly interesting because, for instance, GPT-3 [4] was not explicitly trained for source code generation, whereas the product GitHub Copilot is [5].

Given the considerable success of language generation models in other modalities and the abundance of publicly available code, we hypothesize that analyzing the capabilities of a particular tool in Spanish can help address the challenges of finding evaluation mechanisms [6] and their potential future implementation by companies.

Our study [3] indicated that the most effective models are not available for free use, limiting advancements to what small companies, educational organizations, or even individual programmers can achieve. Their use remains restricted to a few technological companies.

GPT-3, as a model, introduced an innovative aspect: it is defined as a general-purpose (task-agnostic) [7] model, meaning it can be used with minimal fine-tuning for different purposes, such as question-and-answer tasks, extensive text writing, or natural language translation [8].

The launch of ChatGPT [4], which is free to use or available through a very low-cost subscription for the Plus version, based on GPT-3.5 in October 2022, generated unprecedented public interest. Unofficial estimates suggested that by February 2023, there were approximately 100 million users of this tool [9]. This public interest and free access motivated the selection of this tool for our work. Access to ChatGPT is through a browser via a public URL without any particular requirements for its use [4].

1.2 Preliminary Considerations

The anticipation of widespread ChatGPT use and our previous research on the capabilities of Large Language Models (LLMs) [10] led to the planning and execution of this work, where we focus on testing the tool directly from the perspective of end users.

As mentioned in [11], the utility and applicability of ChatGPT for writing code ultimately depend on the level of detail provided by the user when describing the function and parameters of the desired code, and the scale and complexity of the requested code. In this work, we focus on generating standalone functions from plain text prompts, sometimes with examples, and evaluating the automatically generated source code through unit tests.

The documented tests until 2022 mostly used Python. However, in Stack Overflow's surveys of the most popular technologies [12], 2022 marked the tenth consecutive year of JavaScript being the most used programming language, with a 65% usage rate. Therefore, we decided to conduct this study with JavaScript as the target language.

ChatGPT and other GPT models are trained on a large multilingual dataset, which allows them to generate text in multiple languages. However, there is a notable predominance of English in the training data. This predominance results in better performance for English tasks compared to those in other languages, especially when it comes to higher-level tasks like named entity recognition, question answering, and common-sense reasoning [13].

The research question is then: Can ChatGPT produce correct and satisfactory JavaScript source code from descriptions in Spanish?

1.3 Organization of This Paper

This study presents both quantitative and qualitative results, consistent with other studies reviewed in the literature [3]. The structure of this work is as follows:

First, we introduce the problems selected for the study and the categorizations we applied to them (Sect. 2).

Next, we describe the methodology, the tests conducted, and the tools used in the study, along with considerations for specific situations (Sect. 3).

Then, we present the quantitative results, which offer a statistical perspective (Sect. 4).

Following this, we provide a qualitative analysis of the results produced by ChatGPT and the interactions performed (Sect. 5).

Finally, we present our conclusions (Sect. 6) and propose future research directions based on these findings (Sect. 7).

2 Definitions of the Problem or Prompts Dataset

2.1 Considerations

To accurately compare our model, we created a dataset of 250 original programming problems with unit tests. These problems were selected from a set used in recent years by a programming course at our faculty, ensuring they reflect the level and complexity expected of an advanced computer engineering student during an exam.

Most of the reviewed works in [3] evaluate language comprehension, simple algorithms, and basic mathematics, with some comparable to simple software interview questions. To advance further in depth, we decided to avoid simple math problems already shown to be solvable by previous tools.

We focused primarily on finding functions that work with sets, expressed as arrays or matrices, requiring a higher level of abstraction in their understanding and processing.

2.2 Types of Problems Defined

The problems selected for this work involve set and array relationships, such as union, intersection, Cartesian product of arrays or matrices, scalar or vector products, matrix transposition, and even matrix multiplication.

Examples of problems requiring numerical treatment include numerical analysis such as calculating divisors, identifying amicable numbers, perfect numbers, or converters from hexadecimal to decimal, binary to decimal, and vice versa.

We have selected problems that require the analysis of sets of numbers, implemented both as an arrays or matrices, such as calculating medians or biased averages, generating monotonically increasing series, or finding the greatest common divisor of an undetermined size set of numbers.

For more complex problems, we selected implementations for calculating resultants according to Varignon's theorem, computing check digits for car plates, implementing the Vigenère cipher algorithm, distributing seats in elections according to the D'Hondt method, and a variant of the Euler's Königsberg problem that has a solution.

To test the interpretative capabilities of comments, only 75% of the problems were provided with a resolution example (OSL one-shot learning; FSL or few shot learning [14]), leaving 25% solved without guidance examples for input or output (ZSL or zero shot learning [14]). All problem statements and their classifications are available for consultation.

2.3 Problem Classifications

To perform a more thorough analysis of the results, we classified the problems into groups or categories in order to identify patterns that would help us better understand the outcomes.

Data Type. From the perspective of the data the functions need to handle, we categorized the problems into arrays, matrices, text, numbers, and structures.

As tasks involving datasets are more complex and require a higher level of abstraction than those with simple data, these tasks were the most predominant. Specifically, as shown on the following figure, 63% of the data were organized in arrays, and 14% were organized in matrices, making up nearly 80% of the data organized in arrays or matrices. There was also processing of numbers and text, each accounting for approximately 10%, leaving a small portion for structures that emulate classes (from the perspective of programming models) (Fig. 1).

Complexity. Evaluating the complexity of requirements or obtained functions is difficult and arbitrary due to the lack of suitable methodologies. However, based on our teaching experience, we categorize the problems into three categories: low complexity, medium complexity, and high complexity. The Fig. 2 figure shows the distribution.

As in any evaluative process, simple problems are more prevalent in this problem base, with a design ratio of 75/25/5. On the figure above, the categorization considers aspects related to the complexity on the dataset regarding understanding the internal logic or data processing. This translates to actual percentages of 72% for simple problems or problems with low complexity, 22% for medium complexity problems, and 6% for high complexity problems.

Complexity and Type. In the following graph, we analyze the intersection between the problem difficulty and the type of data the function needs to handle (Fig. 3).

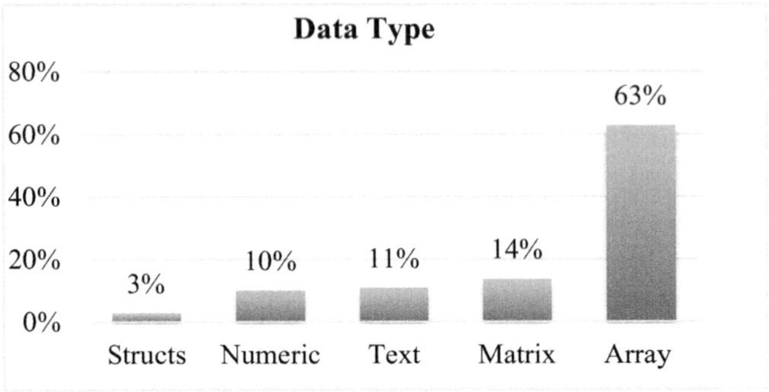

Fig. 1. This figure illustrates the distribution of problems categorized by the type of data they process. The y-axis shows the percentage of problems associated with each data type.

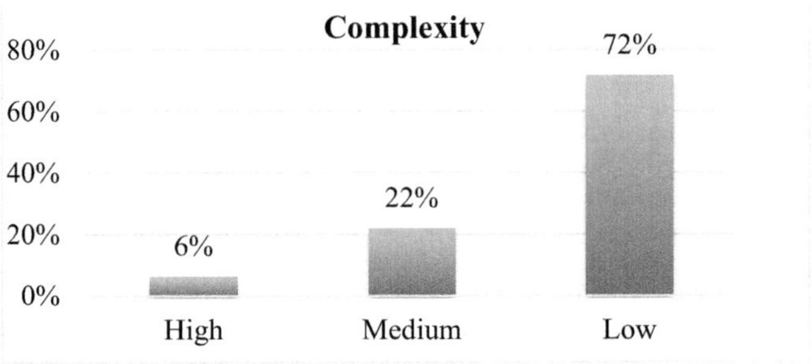

Fig. 2. This figure shows the distribution of problems categorized by their complexity level. The y-axis shows the percentage of problems that fall into each category.

Other Categories: Regarding other analyses of the characteristics of the required functions, we also examined them in terms of the number of input and output parameters. In terms of input parameters, most of the required functions have either one (53%) or two parameters (33%). Combined, these two conditions account for 85% of cases, with the remaining functions having three (7%), four (4%), or zero (3%) input parameters. Regarding output parameters, most functions have one output parameter (75%), with similar distributions for two (8%), three (7%), variable parameters (9%), and four (1%) output parameters.

These characterizations, in terms of the number of parameters, are not as critical in evaluating the difficulty as the complexity of processing. However, we can hypothesize that the more input parameters there are, the higher the resolution difficulty. Similarly, we conclude that functions with more output parameters are also more difficult to resolve. The next figure shows the distribution of complexity over the input and output parameters (Fig. 4).

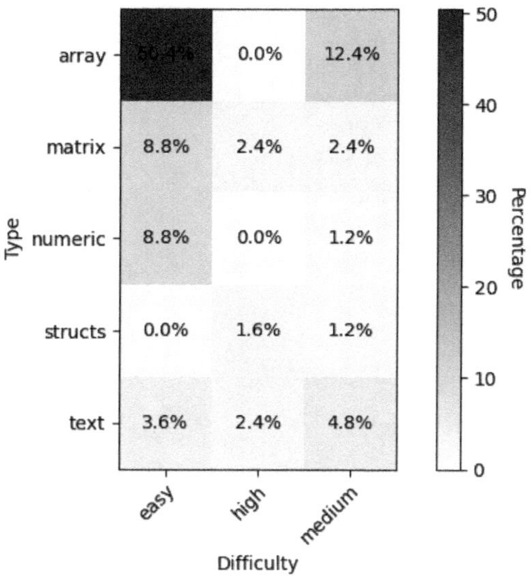

Fig. 3. This heatmap illustrates the classification of problems based on data type and difficulty level. The color intensity and percentages within each cell indicate the proportion of problems that fall into each type-difficulty combination. The heatmap reveals that easy array-based problems are the most common, while more complex tasks involving structs and numeric data are less frequent.

3 Methodology and Evaluation Framework

For code testing, we used NodeJS, a runtime environment for JavaScript built with V8, Chrome's JavaScript engine. We selected Visual Studio Code as the IDE for development, as these tools are free for both private and commercial use, ensuring adoption is not restricted by licensing costs.

All problems were written with the same base, requesting a JavaScript function with a specific, unique name that fulfilled a particular requirement. As previously mentioned, some comments included examples, while others did not.

The first phase of methodology entailed presenting each of the written problem statements to ChatGPT for problem-solving. Below, we offer a couple of examples of the requirements used (Table 1).

The result provided by ChatGPT for the request consists of four parts: an introduction, the source code for the required function, test examples, and a description of the function. Thus, the interaction not only provides the requested algorithm but also offers usage information and aids in understanding, all without the need for explicit requests.

In cases where the requirement included one or more test cases, we observed that ChatGPT typically used them as examples or test cases. For those requirements that did not include examples, the model generated them independently without prior prompting.

We made a copy of the interaction with the model and stored it to have a backup and for detailed analysis. Below, we present some examples of the model's responses to the posed problems.

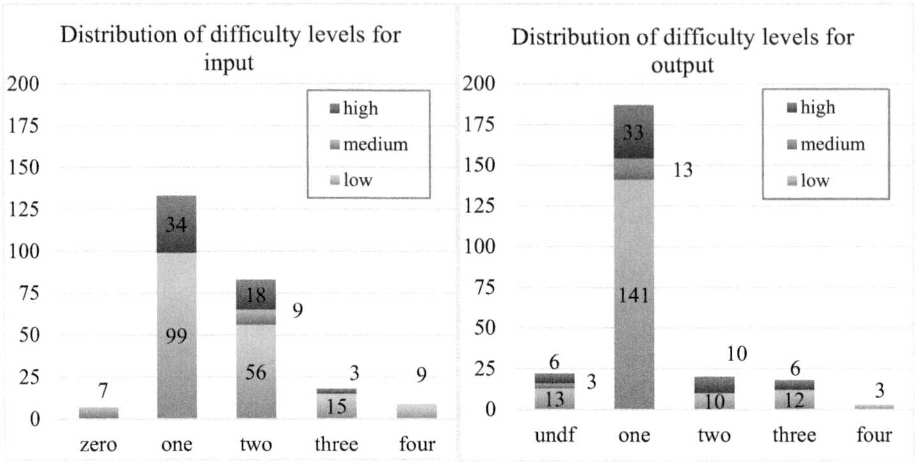

Fig. 4. Classification of Problems Based on the Number of Input and Output Parameters. The figure illustrates the relationship between the number of input and output parameters and the associated difficulty levels for resolving functions. The bar charts are divided into categories based on the number of input parameters (on the left) and output parameters (on the right), with each bar further segmented to show the distribution of difficulty levels. Some requests have zero input parameters, indicating that they work exclusively with internal data. Others have no output parameters, meaning their actions tend towards behaviors like printing on the screen or similar operations. The label "undf" refers to these undefined output cases.

Table 1. Examples of prompts used on ChatGPT

Problem N°	Text
15	Sabiendo que de los meses del año noviembre, abril, junio y septiembre tienen 30 días, febrero 28 días de longitud y los demás meses tienen 31, quisiera una función en javascript que reciba un número para un día (del 1 al 365) y devuelva a que mes pertenece
23	Quisiera una función en javascript llamada mediana que calcula la mediana de un array de números indicados en el primer argumento. Como puede haber dos medianas (en el caso que la lista tenga una cantidad de elementos par), deberá devolver un array sin usar slice

The source code, test examples, and pre-prepared unit tests were integrated into a single file for testing in the IDE with unit tests. We then compiled and executed the function and analyzed the results.

When we refer to compilation, we are referring to the syntax verification and execution of the generated code in a development environment like Node.js. In these environments, JavaScript is executed through a Just-In-Time (JIT) compiler, which compiles parts of the code at runtime to optimize its execution. While there is no prior compilation process as in compiled languages, the JIT compiler performs runtime checks to ensure that the code is syntactically correct and executable.

If the unit tests passed, the problem result was marked as a 1. If the unit test failed, we requested modifications from the tool, similar to how an individual programmer would, to assess its usability from that perspective. This approach contrasts with natural language generation, where samples are typically evaluated heuristically or through automated testing, as explored in our previous study [3]. If a correct result was achieved with a single modification request, the result was marked as a 2.

If a single interaction did not yield a correct result in the unit tests, we examined the code to determine whether the issue was with the request or the implementation. We then attempted to have the model correct the code. If successful within a few interactions, the result was categorized as 3. If more than four attempts were needed, it was categorized as a failure and assigned a 4.

In some cases, the request was to avoid using specific JavaScript functions to simplify the evaluation of the function's development, avoiding functions like splice or slice.

After assigning a score of 1–4 for the functional resolution of the given problems, we reviewed the solutions from a user perspective, focusing on code readability, comments, examples, and the explanation of the function's operation. Previous research has used metrics such as BLEU, and generation metrics have mainly focused on analyzing the accuracy and complexity of the code output rather than its expressiveness and complexity. This qualitative review is used solely as a reference, as it does not appear to be comparable.

4 Quantitative Results

Syntactic Correctness: In all the responses, the source code generated by the model compiled successfully without any syntactic errors, and the test examples provided by the model were suitable for syntactic verification. This represents an improvement over previously reviewed studies.

Functional Correctness: Regarding functional resolution, 81% of the solutions worked correctly directly from the model's output. Cases requiring a single additional interaction with the model accounted for 8%. Those results are shown on the figure below (Fig. 5).

Thus, combining the direct output from the model with those requiring one additional interaction, the success rate was around 90%, which we consider successful based on the parameters determined at the beginning of the study.

Analysis of Correctness Versus Difficulty: Since our classification of simple problems was close to 72%, we analyzed the functional results in relation to the difficulty of the cases. To simplify this analysis, we combined the values for functional results 1 and 2, as these are considered correct, classifying them as GOOD. We combined the values for functional evaluation results 3 and 4, considering them generally incorrect, and classified them as NO GOOD. The next figure shows the crossing the difficulty over functional correctness.

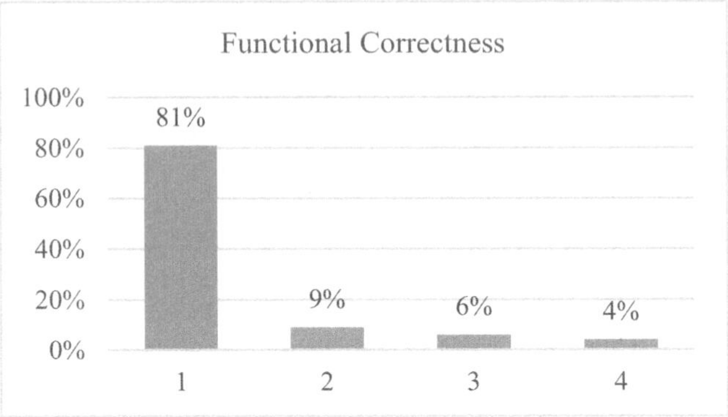

Fig. 5. This figure presents the results of the functional evaluation of the generated functions, categorized by the number of attempts required to achieve a correct solution. The x-axis represents the number of attempts, while the y-axis shows the percentage of cases where the generated function was functionally correct.

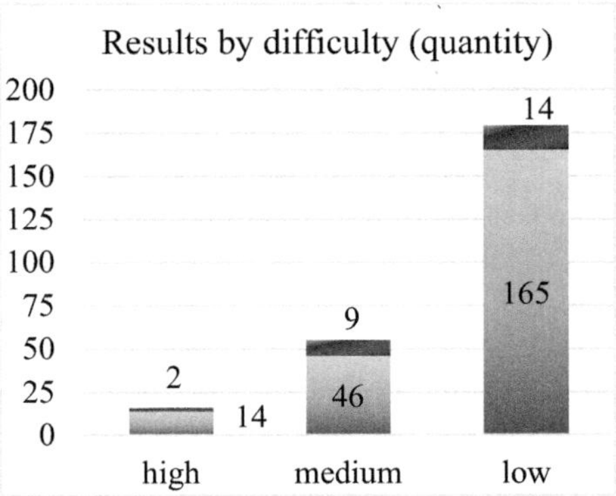

Fig. 6. Illustrates the distribution of successful (GOOD) in green and unsuccessful (NO GOOD) in red outcomes based on the difficulty level of the problems.

As the quantity of samples were dissimilar, they are not comparable on a direct basis. Therefore, we use the percentage in order to try a clearer view in the distribution of successful (GOOD) versus unsuccessful (NO GOOD) outcomes. Figure 6 shows that the number or percentage of successful resolutions for problems classified as easy is higher than for those of medium and high difficulty. The last figure, Fig. 7, illustrates that while success rates are generally high across all difficulty levels, there is a trend of increasing failure rates as the difficulty of the tasks increases.

To verify this hypothesis, we performed a statistical test. A chi-square test was used to determine whether there is a significant association between categorical variables. In the context of analyzing the distribution of GOOD vs. NO GOOD results by difficulty level, the chi-square test would help us evaluate if the observed differences in success rates across varying levels of difficulty are statistically significant, rather than due to random chance (Table 2).

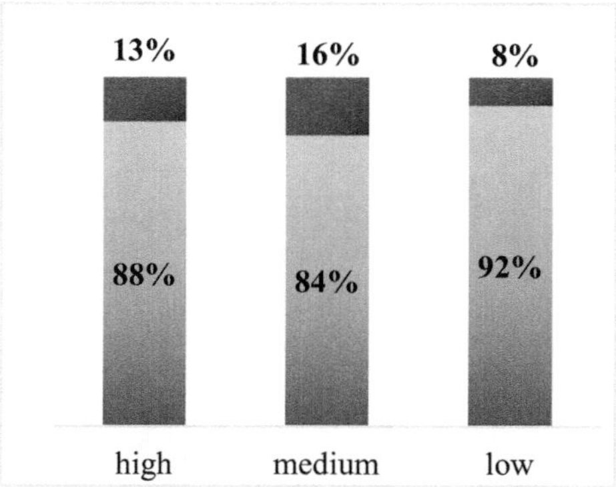

Fig. 7. Illustrates the percentage distribution of successful (GOOD) in green and unsuccessful (NO GOOD) in red outcomes across varying difficulty levels: high, medium and low. The x-axis categorizes the tasks by difficulty, while each bar represents the total proportion of GOOD and NO GOOD results within that category, summing to 100%. This visualization highlights the relationship between task complexity and success rates

Table 2. Chi-squared for results over difficulty.

Variable	Value		
Chi-squared	3.5299926640708756		
P-value	0.17118741439315885		
Degrees of Freedom	2		
Expected frequencies		GOOD	NO GOOD
	High	14.4	1.6
	Medium	49.5	5.5
	Low	161.1	17.9

We observed the test results, and the p-value obtained was 0.1712, indicating that there is not enough evidence to reject the null hypothesis that the success rate is independent of the difficulty level. In practical terms, this suggests that the observed differences in success rates between the different difficulty levels might be due to chance and not to a significant relationship between the problem difficulty and the model's ability to solve them.

Analysis of Correctness Versus Data Type: Using the same parameters for GOOD and NOGOOD, we made the analysis crossing over data type, The following chart displays the functional resolution results in relation to the data types, aiming to identify whether successful resolution depends on the problem's data type (Figs. 8 and 9).

As with the prior analysis, the results by type are not comparable on a direct basis. Therefore, we use the percentage to try to get a clearer view in the distribution of

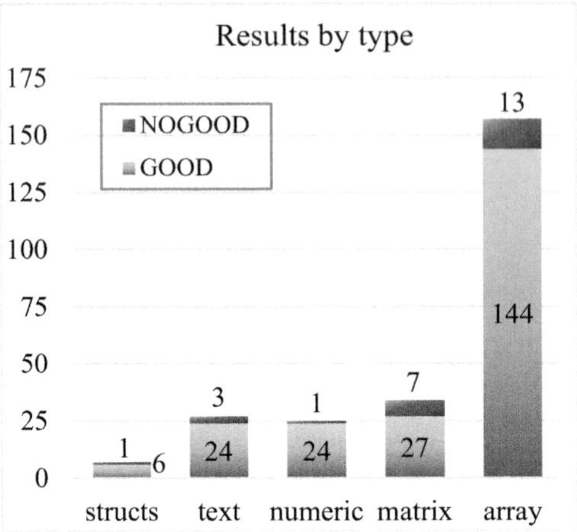

Fig. 8. Shows the distribution of successful (GOOD) and unsuccessful (NO GOOD) outcomes categorized by data type: structs, text, numeric, matrix, and array. Each bar represents the total number of results for a specific data type.

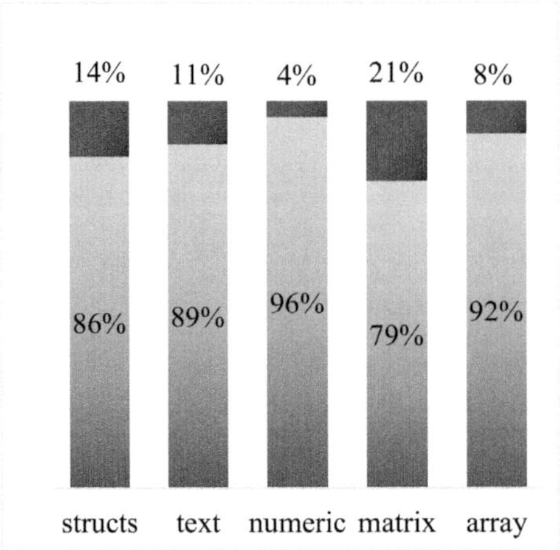

Fig. 9. Shows the percentage distribution of successful (GOOD) and unsuccessful (NO GOOD) outcomes across different data types. Each bar represents the total percentage of results for each data type, with the green section indicating the proportion of GOOD outcomes and the red section representing the NO GOOD outcomes.

successful (GOOD) versus unsuccessful (NO GOOD) outcomes over the type. We also performed a chi-squared test over the data and the categorical values (Table 3).

In this analysis we also get a high p-value, 0.204, indicating that there is not enough evidence to reject the null hypothesis that the success rate is independent of the data types. In practical terms, this suggests that the observed differences in success rates between the various data types might be due to chance and not to a significant relationship between the problem difficulty and the model's ability to solve them.

Table 3. Chi-squared for results over difficulty

Variable	Value		
Chi-squared	5.931111864420219		
P-value	0.20435228707942735		
Degrees of Freedom	4		
Expected frequencies		GOOD	NO GOOD
	Structs	6.3	0.7
	Text	22.5	2.5
	Numeric	24.3	2.7
	Matrix	30.6	3.4
	Array	141.3	15.7

Other Correlations: We also perform tests over input parameters or output parameters, but we couldn't get significance over them.

5 Qualitative Results

The qualitative analysis of the source code generated by the model addressed different dimensions to evaluate the applicability and quality of the code from a programmer's perspective. The dimensions considered for each of the responses to the given problems were as follows.

Clarity and Readability: All the code is highly readable, with proper use of indentation and spacing between blocks, making it very clear when reading. In all cases where the model was allowed to select the names of variables or functions, these were appropriately descriptive. As examples, we can show in Table 4 some results.

Structure and Organization: The code was evaluated from the perspective of organization, variable declaration, and initialization when applicable, and the result was satisfactory. The same conventions in indentation and spacing were respected throughout all interactions.

In some cases, the model generated functions and structures to improve code modularity, and in all cases, it added appropriate usage examples at the end of the source code to test the function.

Explanations and Comments: The source code generated by the model generally lacks embedded comments; however, some definitions do include them. Nevertheless, all functions were accompanied by explanatory text related to their design or usage. These texts often use bold or single quotes to emphasize certain words, improving readability.

Although requesting line-by-line documentation was not part of this study, we recognize that given the explanatory nature of the generated functions, this could be requested, and the model would likely comply. This could be explored in future research.

Table 4. Variable Names

Sample N°	Example
54	const resultado = [...arr]; const totalElementos = arr.length;
56	const vueltoRestante = pago - gasto; const vueltoFinal = [];

80	let maxRatio = 0;
	let partidoGanador = -1;
	for (let j = 0; j < numPartidos; j + +) {
	const votos = resultados.filter(result = > result[0] = = = j + 1).map(result = >
	result[2]).reduce((total, v) = > total + v, 0);
	const ratio = votos / (asignacion[j] + 1);

81	barrios: ['Kneiphof', 'Vorstadt', 'Altstadt', 'Lomse'], const caminoEuleriano = [];

These analytical dimensions could be adapted or expanded depending on the specific objectives of the study. This analysis could be enhanced by conducting expert interviews or content analysis to gather more data on these dimensions, which we will leave for future research endeavors.

It is important to note from this analysis that the model, in some instances, generated more detailed or explanatory results than were explicitly requested.

6 Conclusions

We have observed that the source code generated from all interactions compiles successfully in every case, achieving a 100% success rate. Additionally, we found that the solutions are functionally very close to the requested requirements in 90% of cases. The use of relevant identifier names (variables and functions) related to the problem enhances the readability of the source code.

The primary challenge in automatic code generation is the assumption that the programmer's intent can be precisely expressed through comments and interactions, and that the tool will accurately implement the specified requirements. This raises important questions about which sentences, comments, and syntax are most effective in eliciting the best performance from the model.

The generated code must be tested through unit tests, which can be automated. If the code fails these tests, the programmer must review it and decide whether to correct the code or re-engage with the model. We conclude that even with these tools, and even if the model were perfectly accurate, we would not expect the need for programmers to be eliminated or the labor costs associated with writing source code to be significantly reduced. However, we can foresee an increase in the amount of source code generated per unit of time.

Real-world programming tasks often involve iterative approaches and error corrections, generating alternative samples. Therefore, the failure of a unit test was not considered an error in itself. During the process, we revisited the tool to request corrections or provide a clearer explanation of the intended goal.

These capabilities make ChatGPT a potentially useful tool for code generation and simultaneous explanation. However, it is important to note that while the model can generate useful explanations based on the patterns it has seen in its training data, it does not have a real conceptual understanding of the code as a human would. All its responses are the product of statistical patterns it has learned during training.

Therefore, we suggest that within academic settings, students should be taught to interact constructively with ChatGPT according to the educational institution's ethical guidelines. In professional settings, it would be beneficial for programmers to interact with the tool in a manner that produces source code from comments most effectively and efficiently, provided that intellectual property rights are not violated. This could lead to increased productivity.

Finally, it is important to note that the purpose of this study was centered on interactions and prompts in Spanish. While our analysis is focused on Spanish, the insights gained could be relevant for other non-English languages as well. The findings suggest that similar challenges and benefits might be observed when using ChatGPT in different linguistic contexts, potentially broadening the applicability of our results.

7 Future Research Lines

Given our findings that this model consistently generates syntactically correct source code and is functionally correct in most cases, we believe further research into its impact on the workplace is warranted. Areas for future investigation include:

- Assessing the model's capability to handle entire projects through a single request or multiple interactions, as this study focused on generating individual blocks or functions.
- Exploring the effectiveness of different structured prompts for code generation to determine the optimal way to make requests. This includes investigating whether standardizing verbal forms, as opposed to using more informal natural language, produces better results.
- Evaluating the economic benefits of using a tool to quickly generate complete code blocks that need to be integrated into ongoing projects. This could involve measuring the time it takes for different programmers to write code manually and comparing it with the time saved using this or other model-based tools, considering potential inefficiencies caused by switching development tools.
- Analyzing the impact on software development organizations, particularly in terms of changes to code documentation practices, when the bulk of source code is generated by a tool rather than the programmer. This includes the possibility of the model documenting code blocks and providing line-by-line comments.

References

1. Schatsky, D., Bumb, S.: AI is helping to make better software. AI is helping to make better software. https://www2.deloitte.com/us/en/insights/focus/signals-for-strategists/ai-assisted-software-development.html, Accessed 10 May 2024
2. Zhang, S., et al.: Planning with large language models for code generation. arXiv preprint arXiv:2303.05510 (2023)
3. Bender, A., et al.: Generación Automática de Código Fuente a través de Modelos Preentrenados de Lenguaje, un análisis de la literatura. In: AGRANDA 2022 - Simposio Argentino de Ciencia de Datos y GRANdes DAtos, pp. 50–65 (2022)
4. OpenAI: ChatGPT. ChatGPT. https://openai.com/blog/chatgpt, Accessed 04 May 2024
5. GitHub: Co-pilot. Co-pilot. https://github.com/features/copilot, Accessed 05 Jan 2024
6. Chen, M., et al.: Evaluating large language models trained on code. arXiv preprint arXiv:2107.03374v2 (2021)
7. Finn, C., Abbeel, P., Levine, S.: Model-agnostic meta-learning for fast adaptation of deep networks. arXiv preprint arXiv:1703.03400v3 (2017)
8. Reed, S., et al.: Few-shot autoregressive density estimation: towards learning to learn distributions. In: International Conference on Learning Representations (2018)
9. Hu, K.: Reuters. https://www.reuters.com/technology/chatgpt-sets-record-fastest-growing-user-base-analyst-note-2023-02-01/, Accessed 05 Sep 2023
10. Shanahan, M.: Talking about large language models. arXiv preprint arXiv:2212.03551v5 (2023)
11. Meyer, J.G., Urbanowicz, R.J., Martin, P.C.N.: ChatGPT and large language models in academia: opportunities and challenges. BioData Min. **16**, 20 (2023)
12. StackOverflow: Most popular technologies. Most popular technologies. https://survey.stackoverflow.co/2022/#technology-most-popular-technologies, Accessed 26 May 2023
13. Lai, V.D., et al.: ChatGPT beyond English: towards a comprehensive evaluation of large language models in multilingual learning. arXiv preprint arXiv:2304.05613 (2023)
14. Wang, Y., et al.: Generalizing from a few examples: a survey on few-shot learning. ACM Comput. Surv. **53**(3), 1–34 (2020)
15. Microsoft: visual studio code. Visual studio code. May 2023. https://code.visualstudio.com, Accessed 28 May 2023
16. NodeJS.org: NodeJS. NodeJS. https://nodejs.org/es, Accessed 28 May 2023
17. Papineni, K., et al.: BLEU: a method for automatic evaluation of machine translation. In: Proceedings of the 40th Annual Meeting on Association for Computational Linguistics, pp. 311–318 (2002)
18. Price, M.J.: C# 10 and .NET 6 – modern cross-platform development. Packt Publishing (2021)

Enhancing Live Performances with AI-Driven Visuals: A Machine Learning and Generative AI Approach

Felix Husac and Dana Simian(✉)

Lucian Blaga University of Sibiu, 550024 Sibiu, Romania
{felix.husac,dana.simian}@ulbsibiu.ro

Abstract. Machine Learning (ML) techniques and Generative AI gained an increased importance in the Entertainment Industry, enhancing the audience interest by adding new valences to various entertainment events and performances. The projection of the performance of large LED screens became a common practice for performances made in wide open air or indoor spaces. This article proposes a solution for adding new value to a performance by using, alongside the usual performance projection, an additional different one. This projection considers the actors involved in the performance from different perspectives based on a specific algorithm, including the use of ML techniques and Generative AI. Our approach makes use of existing IMAG (Image Magnification) systems and robotic cameras in a master-slave configuration for real-time tracking of the performance. The images gathered by the robotic cameras are analyzed and refined through separate computer vision processing pipelines. The results of the image processing operations are used to select, composite and display the image of one or many performers against an AI-generated image background. The image generation strategy of the AI model is selected by image sentiment analysis and a threshold-based algorithm. For a better understanding and motivation of our proposal, we include a brief state of the art and comparison of most relevant object detection methods and generative models suitable for our goal.

Keywords: Generative AI · Entertainment Industry · Deep Learning · Image Processing · Computer Vision

1 Introduction

Large venues and public events, such as concerts, theatrical performances, political debates and awards shows often make use of LED walls or video projection methods [7,17,39].

Known as IMAG (Image Magnification) systems in the industry, these systems help ensure that artists, presenters, and graphics on stage are clearly visible to everyone in the audience. IMAG systems consist of cameras and a video

D. Simian and L. F. Stoica (Eds.): MDIS 2024, CCIS 2486, pp. 105–122, 2025.
https://doi.org/10.1007/978-3-031-87386-7_8

switcher. The video switcher allows vision mixer technicians to select different camera angles. The live video feed is then displayed on LED screens or projected onto nearby buildings. These systems enhance audience engagement and improve the overall quality of the event by providing clear visuals for audience members seated farther from the stage. Additionally, IMAG systems can display graphics and information, such as an artist's name, performance title, sponsor logos, or other digital signage or digital art [14,27]. This makes it easier for the audience to connect and resonate with the mood and action of the performers or the overall event. Another advantage of IMAG systems is that they help maintain audience interest and emotional connection to the performance by providing a closer view of the action. Employing an IMAG system also helps event organizers cater to a wider audience, by ensuring that people with visual impairments can get an enlarged, bright image of the performance on stage. A very easy to grasp real life example of this technology in action is a fashion show, where the models are filmed and displayed on large screens as they present the new clothing designs walking down the catwalk. This paper proposes a solution to enhance the experience of classical IMAG systems by integrating generative AI technologies and human detection. Our approach combines live camera feeds with artificially generated imagery that reflects the event's mood and aligns with the emotional responses of the performers. The proposed solution benefits from the real life experience of the first author with live event video broadcasting and IMAG systems. In Fig. 1 is presented a frontal view and a top-down perspective of a suggested scenario for placing the components of our proposed solution. Dynamically changing the background for some of the performance camera angles makes the experience more fun and memorable for the public. The venue and organizers can also benefit from the excitement and engagement created by technological buzzwords such as generative AI, machine learning and human detection, creating a sense of novelty and interest for the event. Alongside with this innovative performance projection, a classical, non-altered projection can be presented to satisfy all types of auditorium. Despite the extensive availability of literature on IMAG systems, including numerous books, manuals, and contributions from industry professionals through online blogs and forums, there is a notable lack of peer-reviewed scientific papers addressing this topic. This indicates that while practical insights, experience and knowledge are widely disseminated among professionals in the field, rigorous academic research has not kept pace. Theater productions and concerts have understood the impact and influence of mood and scene lighting on audience emotions. LED video walls offer a new avenue for stimulating the visual senses of the auditorium [9,37,39]. IMAG systems are frequently used at festivals, sports events and performances for displaying art, motion graphics and camera angles of the action, making them a versatile and indispensable tool in the toolkit of event and venue organizers. In classical video and film production, advancements in computational power and LED video wall technology have given birth to a new paradigm of blending CGI (Computer Generated Imagery) with real elements. LED virtual production volumes are used to combine live footage with computer graphics directly in camera. Environmental information, such as camera angles in real life, are applied to digital elements inside graphics engines and then projected on the volume displays [40,49,54].

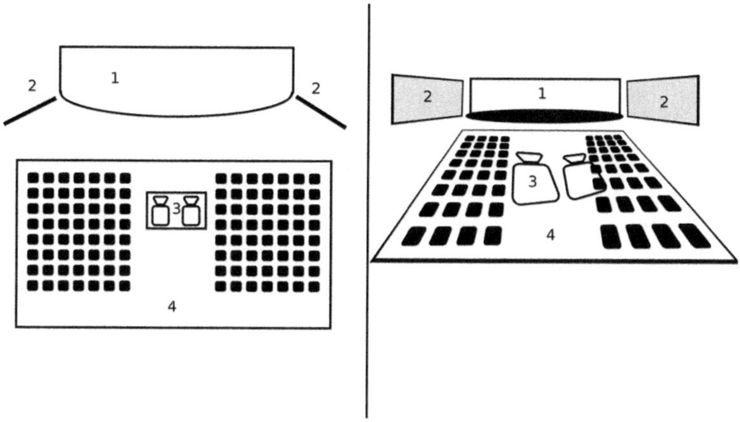

Fig. 1. Simplified, Top-Down and Frontal view of a stage venue. Legend: 1-Stage; 2-LED video walls -for IMAG or other visual media; 3-PTZ Cameras position; 4-Audience.

The live event production entertainment industry can benefit greatly from combining and incorporating such emergent technologies into its workflows. This paper describes such a possible workflow integration. Section 2 describes the theoretical framework of our proposal, explaining the core concepts of our approach. The third section details the methodology employed, from image acquisition to prompt engineering, while the fourth section presents a state-of-the-art overview of machine vision and deep learning generative models necessary for our use case. The proposed selection strategies for model and subject selection processes are presented in Sect. 5. Future research avenues for this article are outlined in the next chapter, along with discussions about the ethical and legal implications of human and facial detection. A conclusion section closes this paper.

2 Theoretical Framework

2.1 Conceptual Model

We present a theoretical framework aimed at improving existing IMAG installations with minimal modifications, particularly for public events such as concerts, fashion shows, and theater performances that feature multiple performers on stage. Our goal is to enhance the viewer experience, event value and crowd cohesion by employing a second LED video wall or projection surface that displays real time video of the performance combined with artificially generated backgrounds. The COVID-19 pandemic significantly accelerated the adoption and evolution of video conferencing platforms, transforming them from niche products into daily communication tools [16]. Prior to the pandemic, Zoom had approximately 10 million daily users, a number that skyrocketed to over 300 million users by April 2020 - according to Statista -, illustrating the shift towards remote interactions [51]. The concept for background blurring and background replacement in video conferencing apps existed prior to the pandemic, but gained significant traction during its period. Automatic background blurring was particularly appealing for users, since it provided a way for them to create a more controlled environment, reduce distraction and enhance privacy. The possibility to include customizable backgrounds allowed users to personalize their surroundings and to express their identity or brand visually, which further contributed to a more professional image they wanted to project in a virtual context. The proposed solution works in a similar fashion, by selecting the desired subject from inside the master camera's field of view and acquiring a zoomed in shot from the slave camera of the same subject. The clean image from the salve camera is then processed to eliminate the other elements of the image, making them transparent, keeping only the subject in the final frame. The generative algorithm is fed a parametrized prompt from which a custom, unique image is created and used as the background for the video stream of the subject. The images are then combined and sent to the video mixer of the IMAG system to be projected on screen. In our solution, background images are obtained via a parametrized prompt - a structured input or instruction, given to an AI model in order to guide its output. The prompt is a description of the desired content that should be found in the output of the generative model. Effective prompt engineering is crucial for generating relevant and high-quality images. This requires clarity, specificity, context, and articulation of both desired and negative traits. The prompts for background generation are obtained based on a multi-parameter strategy. Some of these parameters can be defined as apriori conditions (e.g. intended fantasy locations chosen in advance by the event organizers, seasons, time of the day, the mood and type of the event) while others result from real-time measurements (e.g. the emotional response of the audience or performers). The chosen generative model should be pre trained based on a previously compiled dataset curated before the show begins, by the event organizers. This is to safeguard against hallucinations and problematic images. Our proposed system also employs human supervision during both camera shot selection and background generation.

2.2 Key Components

The approach presented in this article materializes in a system with two components: a hardware and a software component, both described in Sect. 3. The background generation process and image processing pipeline are governed by a set of rules and parameters that are outlined in later sections. To capture live performance, we make use of two robotic cameras that are linked in a master-slave configuration. They autonomously sweep over the performance area and select close-up shots of performers using facial and emotion detection algorithms.

In situations where a large number of performers are on stage (e.g. opera performances with large orchestras, choirs, large theater productions, etc.), projecting all of them can be impractical. The subject selection process is greatly influenced by the result of emotion analysis algorithms. In such cases, an average emotional response score will be calculated for all the individuals present within the frame, and an individual will be selected at random. The close-up camera angle of the chosen subject is processed to isolate the performer from the original background. The image generation pipeline then creates an appropriate artificial background based on preconditions and contextual information. The processed video feed and generated background are composited together and sent to the vision mixer. The desired result is a close-up shot of a performer, composited on an artificially generated image, and displayed on a secondary IMAG video wall. Compositing is the process of layering or blending together two or more visual elements to form a single final image. The process of compositing originated in film and television production. It was first introduced to blend visual effects (VFX) with live-action photography. Initially reliant on techniques like rotoscoping, where subjects were manually isolated from their backgrounds, compositing has evolved with digital methods such as chroma keying, which removes solid color backgrounds to create transparency channels [1,40]. Advancements in computer vision and machine learning now allow for real-time image segmentation and subject isolation. Notable day to day examples of this technology in use are the user customizable backgrounds offered by video conferencing applications, such as Zoom or Google Meet [16].

3 Methodology

The image acquisition pipeline is conceived in such a way as to be both cost effective to implement, and integrate with existing IMAG systems without compromising quality and ease of use. The system we propose is composed of two main components: a hardware component, and a software component.

3.1 Hardware Setup

The hardware component consists of two robotic PTZ (Pan-Tilt-Zoom) cameras capable of autonomously or remotely changing their orientation in the X and Y axes (pan and tilt), and field of view in the Z axis (zoom). The cameras are to be connected in a master-slave configuration, as presented in Fig. 2. PTZ cameras are commonly used in remote areas, where the presence of a human operator is practically impossible due to limited space, inaccessibility (if the camera is placed high up on the stage's steel frame) or where the presence of a bigger camera, tripod and operator is non aesthetical, such as directly on the stage in front of the performers. They are also employed in the field of surveillance systems. Such cameras can be operated remotely from inside the video control room, or can be set up to automatically follow a subject by employing built-in AI tracking features, if the camera models used have this feature implemented. Accurate subpixel level encoding of the movement axes is needed for a satisfactory tracking result [22,25,33]. Robotic camera gimbals, with servo-powered zoom and servo-powered focus are also available for cameras that are not natively designed to move and change focal distance without the input of a human operator, making the system versatile for any sort of camera the venue wishes to use for the master-slave configuration. The camera systems are connected to a powerful computer that runs a target selection and tracking algorithm, and a feature detection, image segmentation, background generation and image compositing pipeline. The output of the system is fed in real time to the existing IMAG infrastructure of the event venue. The master camera is tasked with exploring the space of the performance area, depending on stage size and number of performers, while the slave camera exploits the subjects found in the field of view of the master camera. The slave camera follows the movements of the master camera and zooms in on potential target candidates. Potential subjects are evaluated by the computer, from the pictures acquired by the master camera.

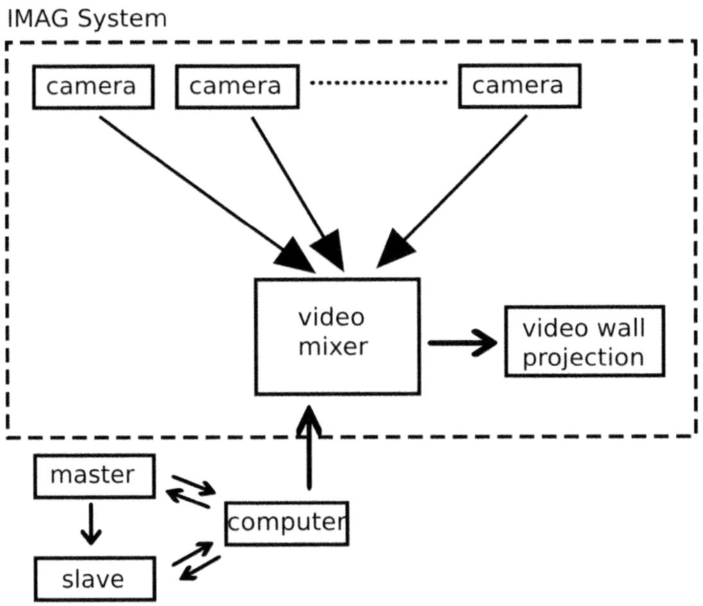

Fig. 2. High level overview of how the hardware component of our solution integrates with a typical IMAG system

3.2 Image Processing Pipelines

Image processing pipelines consist of the Image Acquisition and Compositing Pipeline and the Image Generation Pipeline.

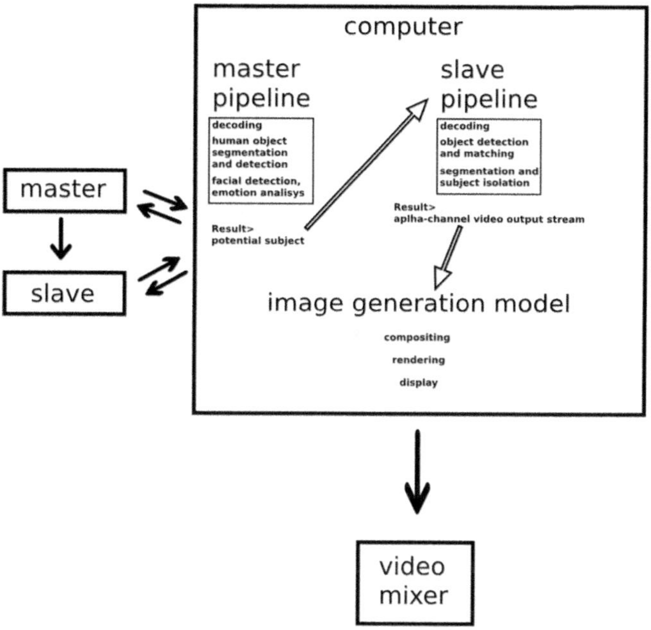

Fig. 3. Visual representation of the complete image workflow between the two cameras.

Image Acquisition and Compositing Pipeline. The images captured by the two cameras are processed independently, each with distinct objectives, using a black box model. The master camera transmits images of the stage or performance sector it is aimed at to a targeting algorithm. A feature detection algorithm identifies humans and their faces in these images, generating a list of potential candidates for the slave camera to focus on. The candidate exhibiting desirable traits, such as an agreeable emotional state, is selected based on a pre-established targeting algorithm, allowing the slave camera to zoom in on this subject. The slave camera's video feed is also processed independently, employing feature detection and segmentation algorithms to isolate the chosen candidate. This processed feed, which features only the subject against a removed background, can be sent to the vision mixer or video processing program as the key source. Meanwhile, an AI-generated background serves as the fill source, transmitted separately. An alpha matte, functioning like a binary mask, defines transparency levels for this key source. The key/fill process is a standard compositing method used in television production to combine dynamic footage and graphics in real-time, most notably used in chroma keying. The alpha matte acts as a transparency map, indicating which areas of the fill layer are visible by keying out unwanted portions based on their transparency values. This ensures a seamless blend between the key and fill layers, ensuring that only the intended elements are visible in the final composite video [56]. Compositing video elements directly within the computer streamlines processing by reducing the load on the video mixer. This lowers complexity for the video mixer operator and frees up inputs for additional camera angles, instead of having multiple connections taken up by different elements of the final picture. Such results can be achieved using image processing libraries such as OpenCV, or dedicated video streaming software, paid or open source, like vMix or OBS Studio, which support real-time image compositing. The master pipeline, from Fig. 3 analyzes a larger frame and comes up with a potential subject on which the slave camera zooms in, separates it from the background and sends the alpha transparency video to be composited, rendered and sent to the vision mixer.

Image Generation Pipeline. The generative models employed in this application need to operate in real-time or near real time, in order to keep up with the video feed. They need to generate the image as soon as a subject is selected and the prompt is assembled, in order to ensure the selected individual maintains their emotional state, body pose and mood. Selecting the right generative model, compiling a training dataset and refining the model is a process that takes time and requires attention to detail, with a focus on the end goal. There are two possible approaches for streamlining this process and making it easier to use by event organizers: using online, cloud based, pretrained generators, or using a local generative model. One notable cloud generative model is Leonardo.ai [29]. The feature that distinguishes it from other competitors is that it offers its users the possibility to add their own dataset and train the model on their desired pictures. [30] Several real time or near real time local image generators are presented in the state-of-the-art section of this paper. Selecting the right generative model for the desired use case is a function of speed, accuracy and desired stylistic or artistic results that can be explored by the process of empirical testing.

3.3 Prompt Engineering

A key step for image generation is prompt engineering [53]. Parametrizing the prompt is a key step for guiding the AI model effectively. Defining clear parameters and establishing event conditions are steps that need to take place before the live event begins. The organizers need to take into account the mood of the event, the target audience, and other stylistic and artistic decisions that will guide the image generation process. After establishing the pre-event conditions, the specific parameters that will be included in building the prompt are clearly defined, allowing information that will be collected at the image analysis step to influence the output of the chosen generative model. After selecting the target individual, the prompt is constructed by combining the given parameters with the chosen preconditions. The prompt needs to include a character description, asking the model to generate a suitable, matching background to the visual aspect of the person, and it needs to specify contextual information regarding mood, setting and atmosphere (e.g. during sunset, at the beach etc.), based on the other parameters gathered during the frame analysis step. The emotional tone of the prompt should be set by the result of the emotion detection step. Constraints or negative inputs should be considered beforehand, in order to steer the generation process. Before reaching the generative model, the prompt runs through a preconditioned LLM (Large Language Model). The conditions set upon the language model match the pre-event conditions selected by the organizers. This step ensures that the prompt will include more descriptive keywords, making it more detailed, and will be formatted in the optimal way for the generative model to operate with (including model-specific conditions, image style etc.). During this step, the user is allowed to override or adjust the prompt before being fed to the image generation model. These steps ensure that the prompt maintains contextual awareness of the event conditions and the real time gathered parameters. This way of building the prompt ensures that both the chosen individual and their generated background are contextually appropriate and emotionally resonant. This also ensures the generation of high-quality images that follow the set conditions, while maintaining diversity and relevance.

4 State of the Art Object Segmentation and Extraction

Object detection focuses on identifying and localizing specific objects within an image by assigning bounding boxes, without providing pixel-level detail about the objects' shapes or boundaries. Segmentation goes a step further, by splitting the image into semantically meaningful zones and providing object boundaries. Detection and segmentation can be thought of as complimentary operations, depending on their use case and desired result. In computer vision and image processing, image segmentation and feature extraction are important techniques, particularly in the context of human detection. This section provides an overview of the fundamental concepts, methodologies and techniques used in image segmentation and feature extraction, focusing on their application to human detection.

4.1 Image Segmentation

The process of partitioning an image into meaningful regions is called image segmentation [18]. Segmentation is an image processing operation that facilitates the further analysis and visual understanding of image data. It can be classified into three primary categories: semantic, instance and panoptic segmentation [4]. Semantic segmentation assigns a class label to each pixel in an image, without distinguishing between different instances of the same object type. Instance segmentation compliments semantic segmentation by labeling both individual pixels and differentiating between instances of the same class, permitting the identification of multiple objects of the same category in an image. Panoptic segmentation combines the two previous methods [4,24]. It isolates the meaningful parts of the image, assigns a unique identifier to each object alongside a semantic label that classifies the objects into distinct categories [24]. Panoptic segmentation is most efficient in use cases such as autonomous vehicles, medical imaging and augmented reality. There are many techniques a segmentation operation may employ in order to achieve its goal of separating objects: thresholding is the most basic method of segmentation, dividing pixels based on intensity values in relation to a given threshold; edge-based segmentation focuses on identifying boundaries by sudden changes in intensity or color; region segmentation groups pixels into regions based on given similarity criteria; clustering has a similar behavior to region segmentation, utilizing feature vectors as a basis for the clustering algorithms; and deep learning approaches use techniques like CNNs (Convolutional Neural Networks) to automatically learn features from data.

4.2 Feature Extraction

Feature extraction is the process of identifying and quantifying relevant attributes or characteristics in an image. It is commonly used in classification tasks, and is a preliminary step in more complex use cases like object detection [35]. The process of detecting an object consists of three steps: segmentation as described above, attribute computation for each segment and classification of the features into specific object categories. There are several approaches one can use to implement object detection functionality: one-stage detection and two-stage detection approaches. Single stage detection is optimized for speed, and simultaneously performs detection and classification, being implemented in models such as YOLO (You Only Look Once) [43] or SSD (Single Shot Multibox Detector) [34]. Two-stage approaches propose potential object regions and then classify them. They are implemented in R-CNN (Region-based Convolutional Neural Networks) [11] and its variants, and FPN (Feature Pyramid Network) [31]. For the human detection and facial detection tasks of our software component, instance segmentation is the most suited approach, because it is capable of counting and tracking multiple individuals at the same time, while semantic segmentation only offers information about the overall presence of human subjects inside the frame, without differentiating between individuals. Panoptic segmentation offers both contextual understanding of the frame and individual instance differentiation, but for the use cases described in this paper, only the number of individuals and the attributes of each object - height, width, shape - are needed and extracted from the segmented human figures. R-CNN and its variants (Mask R-CNN [13], Fast R-CNN [12], Faster R-CNN [44]) are an advanced method that combines object detection with semantic segmentation, providing pixel-level masks for detected objects. Mask R-CNN, an extension of Fast R-CNN, is capable of outputting the cutout of the object of interest with transparency in the rest of the image due to its pixel accurate masking. MobileNet is a lightweight CNN (Convolutional Neural Network) class architecture, developed by Google, designed for efficient image processing tasks, particularly suited for real-time mobile and embedded vision applications [15]. It provides real-time performance while maintaining a compact model size, based on depth-wise separable convolutions, significantly reducing computational complexity and maintaining comparable accuracy to more complex models. Developers may choose to implement their own custom segmentation models tailored to their specific needs, or choose to incorporate existing, pre-trained deep learning models and libraries like BodyPix and Mediapipe [16]. There are several object detection and segmentation models that make themselves remarkable on the COCO2017 (Common Objects in Context) dataset [6]. YOLOv4 [3] improves on the YOLOv3 [42] algorithm, which could only process around 45FPS (frames per second) at low resolution, now operating at around 65 FPS on Tesla V100 GPUs, with an AP (Average Precision) of 43.5. The latest version, YOLOv8, archives an impressive speed of approximately 40 FPS (25 ms/image) at a mAP (Mean Average Precision) of 55.2, at an IoU (Intersection over Union) threshold of 0.5 [55]. For mobile applications or lightweight solutions, MobileNet V2 offers a speed of 7 FPS to 13 FPS (measured on a Google Pixel 1 phone), with a mAP score of 22.1 [46]. It is more suited for applications with limited computing power. RetinaNet and EfficientDet are slower than the above algorithms. RetinaNet is a single-stage detector, capable of operating at approximately 5FPS, with an AP of 39.1 [32]. EfficientDet is slower, but offers greater precision, with an AP of 55.1 [52]. The two-stage models like Faster R-CNN and Mask R-CNN prioritize accuracy over speed, but are generally slower than their counterparts, running at about 200 ms (5 FPS) on a single GPU, with Mask R-CNN being slightly faster than its counterpart [13]. Speed metrics are highly influenced by the computing power of the graphics cards used in the hardware. The comparison between these various object detection models focuses on human and facial detection. The Average Precision score was calculated based on the IoU metric. IoU measures the overlap between two bounding boxes: the ground truth box, and the predicted bounding box. The higher the IoU threshold, the higher the accuracy of the prediction, but this metric plateaus if the bounding boxes do not overlap [45]. The values for speed and accuracy may vary in real life applications due to differences based on hardware configurations and specific implementations. FPS values over 30 frames per second are desirable, due to the standard television frame rate of 29.97 FPS in NTSC and 25 FPS in PAL systems respectively.

4.3 Generative Models in Entertainment

A notable example of a real-time capable generative model is the state-of-the-art FlowTurbo model, capable of generating images faster than the duration of a 25 FPS video frame. One video frame at 25 frames per second is 40 ms, and FlowTurbo can achieve 38 ms per image, with a FID score of 3.93 [58]. JoliGEN is a framework capable of real time generation. It is an integrated framework for training custom generative models, supporting both GANs (Generative Adversarial Networks) and Diffusion models. Its designed uses are controlled image generation and augmented reality [20,21]. FLUX.1, a rectified flow transformer, was also introduced in 2024, and offers a suite of models designed for text-to-image synthesis. It is also capable of real time image generation, and offers different variants that can be tailored to fit specific performance needs, style and diversity in the generated images. It outperforms competitors on several key ranking parameters, such as visual quality, scene complexity, output diversity and prompt following to name a few [2]. Openjourney and Stable Diffusion (SDXL Turbo) are older models, introduced in 2023. They are both capable of real-time or near real-time image generation. Openjourney is an open-source model inspired by Midjourney. It's standout feature is that it uses a large dataset of artificially generated images for training. It was trained on the Stable Diffusion architecture, using over 124k images, generated by Midjourney AI [41].

SDXL Turbo is a powerful model that employs ADD (Advresarial Diffusion Distilation). This allows the model to offer single-step image generation, allowing for faster outputs [47,50].

While the Frechet Inception Distance (FID) or Inception Score (IS) metrics - used to assess the quality and diversity of artificially generated images - are provided for some models, others may not have a publicly available score due to their recent introduction or ongoing evaluation.

5 Proposed Implementation

5.1 Candidate Selection

Both image processing pipelines implement similar machine vision techniques, but with different goals in mind. The master camera video feed is used to determine the number of subjects in the frame, detect their faces and pass the facial information of the subjects to an image emotion analysis algorithm. The emotion analysis step will identify the most likely candidate to be shown next on screen, in accordance to predefined thresholds and desired target emotions. This step has great influence over the output of the system, both at the candidate selection level, and at the generative level. If the subject displays the correct emotional response - e.g. happyness, melancholy etc. - above a certain threshold of confidence determined by the deep learning model analyzing the facial expressions, then the slave camera will focus on the selected target. The subject's emotions may also influence the output of the background generative algorithm, due to it being included in the prompt. Candidate selection can employ techniques and steps traditionally found in refining genetic algorithms, in order to come up with the best possible fit. Some important mentions are Roulette Wheel Selection, Tournament Selection and Elitism [19,26].

Roulette Wheel Selection favors candidates that align closely with the desired attributes (e.g. facial expression, age, gender, ethnicity, color of clothes etc.), by ranking individuals based on their fitness score, but may struggle to offer diversity of individuals. Tournament Selection is most suited to the scanning approach the master camera takes when surveying a large stage. It randomly selects a subset of individuals and chooses the fittest one. A balance between speed and diversity of selection can be maintained by adjusting the size of the tournament, in this case the zoom value of the camera lens, varying the number of candidates in the field of view of the camera. Elitism can be employed when the number of artists is smaller, or when the stage has set dressings that obstruct the view, which makes exploration of new areas more difficult. This technique preserves good candidates between selections, making sure that a number of the fittest candidates are carried over to the next selected generation. The steps for selecting a suitable candidate are as follows: initializing a diverse population of candidates by rotating the camera around the stage; selecting a frame that represents the current generation of candidates; evaluating each candidate within the frame based on the chosen fitness function (a weighted scoring function for the previously specified attributes); skipping the crossover and mutation steps; and repeating the evaluation process, as local convergence criteria are sufficient. The slave camera feed is a zoomed in copy of the master camera, closely framing the subject identified by the sentiment analysis step of the master camera's image processing pipeline, with the help of the visible features and pattern matching image processing algorithms. The desired result of the slave camera image processing pipeline is a clean video feed that isolates the subject from their background, delivering images with an alpha transparency channel for the subject of interest.

5.2 Model Selection

It is important to prioritize generative models that have high timewise performance. Lower processing times mean a faster time to output, ensuring that the compositing and display steps can take place quickly, while the selected subject is still displaying the traits that influenced their selection. Models can be selected by following a simple algorithm. Firstly, selection criterias need to be established. They depend on the event type and the event preconditions. The selection criterias will always feature the following parameters: real time generation capability, image quality, image diversity and model flexibility; ensuring a consistent image result regardless of the selected model. The possible candidates (e.g. GANs, VAEs, Diffusion models, ADDs etc.) are then selected based on the output of a weighted function that takes into account the importance of the selection criteria and the value of the model's metrics. This process is repeated if the results generated after using the resulting choice are unsatisfactory. It is important to note that each type of generative model greatly influences the output due to its inner structure and the datasets it was trained on. User input variability and real time data integration also have a significant impact on the resulting image. By applying this algorithm to select an image generation model, the user can systematically explore various architectures while optimizing for critical performance metrics.

6 Discussions

Similar to large language models, text-to-image generation models suffer from contextual ambiguity in prompts [23,36]. This ambiguity arises from the inherent complexities of natural language and the model's interpretation of prompts: linguistic ambiguity can lead to varied outputs that may not align with user expectations; words and phrases that have different meanings depending on context can confuse the model, leading to semantic ambiguity; cultural and contextual nuances that humans naturally interpret may prove difficult to understand for generative models, resulting in images that misrepresent the intended meaning of the prompt [57]. On the other hand, complex multi-component prompts may hinder context awareness. Models may still produce outputs that do not fully capture user intentions due to their reliance on statistical patterns learned from training data rather than true comprehension of context. These issues are known as alignment and intent. Ongoing research aims to improve generative output to closer reflect the user's vision through different strategies, while acknowledging the limitations inherent in current AI technologies [5,28].

An important aspect to consider when deploying the proposed application described in this paper is compliance to ethical generative AI use, facial recognition technology regulations, data protection and copyright standards. In a broader sense, GDPR policies restrict the use of personally identifiable information, but copyright laws entitle people to copyright to their own likeness. Even where consent is given for the use of such technologies, the authorities may still rule against it if the technology is deployed from a position of power over the subjects. EU regulations and legislation proposals mainly target biometric identification use cases, such as law enforcement, and classify FRTs (Facial Recognition Technologies) into two main categories: high risk (mass surveillance applications, identification systems) and low risk (for categorisation purposes). High risk systems are to be highly regulated and supervised, while low-risk applications need to be compliant with transparency of information policies [8,10].

By limiting the use of the technology described in this paper to only the artists on stage, event organizers can avoid most legal concerns, due to the fact that artists and performers are considered to be public figures [38].

Illegally sourced training datasets, including those acquired by scraping social media content without consent or utilizing images not in the public domain, without compensating their creators, present an ethical dilemma for generative AI. In response, some artists have taken to contaminating their artworks with image elements specifically crafted to deceive the model during its training process. They use specific tools available online for this purpose, such as Nightshade. The process of intentionally corrupting the training data of a deep learning model, in order to compromise the model's output integrity is called "dataset poisoning" [38,48]. Implementing this research may involve some ethical implications in regards to human behavior and suggestibility, the influence human emotions and facial expressions have on viewers, and raises questions about the degree of influence the event organizers have over the crowd, and if a bad actor could potentially maliciously influence the mood of the event.

7 Conclusions

This paper presents an innovative approach to using existing IMAG systems at venues and public events, such as concerts and theatrical productions, to enhance audience engagement, attention, and the emotional atmosphere of the event. We believe this solution improves the mood, fun factor, and technological novelty of the experience, benefiting both organizers and attendees. Our approach integrates robotic PTZ cameras, machine vision, and generative algorithms into a pipeline compatible with current IMAG systems, ensuring a seamless experience for both audiences and operators.

The proposed system comprises hardware and software components, providing an automated method for selecting and enhancing camera angles with AI-generated image backgrounds. The software includes two image processing pipelines: one for object detection and image segmentation, and another for creating artificial backgrounds. We provide an overview of suitable machine vision algorithms and techniques for object detection and segmentation, as well as real-time image generation models. Additionally, we address ethical considerations and the fair use of generative AI in line with copyright laws and privacy regulations.

References

1. Ahad, M.A.R., Mahbub, U., Turk, M., Hartley, R.: Computer Vision: Challenges, Trends, and Opportunities. 1st edn. Chapman and Hall/CRC (2024). https://doi.org/10.1201/9781003328957
2. BlackForestLabs: Official Product page, FLUX.1, 2. Accessed 02 Dec 2024
3. Bochkovskiy, A., Wang, C., Liao, H.M.: YOLOv4: optimal speed and accuracy of object detection. arXiv, abs/2004.10934 (2020)
4. Borse, S., Park, H., Cai, H., Das, D., Garrepalli, R., Porikli, F.M.: Panoptic, instance and semantic relations: a relational context encoder to enhance panoptic segmentation. In: Proceedings of the 2022 IEEE/CVF Conference on Computer Vision and Pattern Recognition (CVPR), pp. 1259–1269 (2022)
5. Cheng, Y., et al.: Bridging the intent gap: knowledge-enhanced visual generation. arXiv, abs/2405.12538 (2024)
6. COCO Common Objects in Context: Official Dataset Page, cocodataset.org. Accessed 02 Dec 2024
7. Ekim, B.: A video projection mapping conceptual design and application: YEKPARE. Turk. Online J. Des. Art Commun. **1**(1), 10–19 (2011)
8. Madiega, T., Mildebrath, H.: European Parliament: Directorate-General for Parliamentary Research Services, Regulating facial recognition in the EU - In-depth analysis. European Parliament (2021). https://doi.org/10.2861/140928
9. Frank, L.: Real-Time Video Content for Virtual Production & Live Entertainment: A Learning Roadmap for an Evolving Practice. 1st edn. Focal Press, New York (2022). https://doi.org/10.4324/9781003206491
10. GDPR Advisor, GDPR and Facial Recognition: Privacy Implications and Legal Considerations. https://www.gdpr-advisor.com/. Accessed 02 Dec 2024
11. Girshick, R., Donahue, J., Darrell, T., Malik, J.: Rich feature hierarchies for accurate object detection and semantic segmentation. In: Proceedings of the 2014 IEEE Conference on Computer Vision and Pattern Recognition, Columbus, pp. 580–587 (2014). https://doi.org/10.1109/CVPR.2014.81
12. Girshick, R.: Fast R-CNN. In: Proceedings of the 2015 IEEE International Conference on Computer Vision (ICCV), Santiago, pp. 1440–1448 (2015). https://doi.org/10.1109/ICCV.2015.169
13. He, K., Gkioxari, G., Dollár, P., Girshick, R.: Mask R-CNN. In: Proceedings of the 2017 IEEE International Conference on Computer Vision (ICCV), Venice, pp. 2980–2988 (2017). https://doi.org/10.1109/ICCV.2017.322
14. Hot Electronics Co. Ltd.: The Impact of LED Screens on Immersive Entertainment Experiences, Product blog page (2024). Accessed 02 Dec 2024

15. Howard, A., et al.: MobileNets: efficient convolutional neural networks for mobile vision applications. arXiv, abs/1704.04861 (2017)
16. Hrib, E.N., Alboaie, L.: Removing individuals from video streams through facial recognition. In: Proceedings of the 2023 22nd RoEduNet Conference: Networking in Education and Research (RoEduNet), Craiova, pp. 1–4 (2023). https://doi.org/10.1109/RoEduNet60162.2023.10274914
17. Hu, Y.: LED digital image design and production in large stage performance-take "shanghai science and technology festiva" series of performances for example. Art Des. Rev. **6**, 148–159 (2018). https://doi.org/10.4236/adr.2018.63015
18. Hussain Dar, N.: Image Segmentation Techniques and Its Applications (2020). https://www.researchgate.net/publication/340087951_Image_segmentation_Techniques_and_its_application. Accessed 02 Dec 2024
19. Jebari, K.: Selection methods for genetic algorithms. Int. J. Emerg. Sci. **3**, 333–344 (2013)
20. JoliBrain: joliGEN, Official GitHub repository. https://github.com/jolibrain/joliGEN. Accessed 02 Dec 2024
21. joliGEN: Official Documentation Page. https://www.joligen.com/doc/. Accessed 02 Dec 2024
22. Kang, S., Paik, J.K., Koschan, A., Abidi, B.R., Abidi, M.A.: Real-time video tracking using PTZ cameras. In: Proceedings of the Sixth International Conference on Quality Control by Artificial Vision, vol. 5132, SPIE (2003). https://doi.org/10.1117/12.514945
23. Keluskar, A., Bhattacharjee, A., Liu, H.: Do LLMs understand ambiguity in text? A case study in open-world question answering. arXiv, abs/2411.12395 (2024)
24. Kirillov, A., He, K., Girshick, R.B., Rother, C., Dollár, P.: Panoptic segmentation. In: Proceedings of the 2019 IEEE/CVF Conference on Computer Vision and Pattern Recognition (CVPR), pp. 9396–9405 (2019)
25. Larsen, M.V., Mathiassen, K.: Achieving sub-pixel platform accuracy with pan-tilt-zoom cameras in uncertain times. IEEE Trans. Rob. (2024). https://doi.org/10.1109/TRO.2024.3508141
26. Lata, S., Yadav, S., Sohal, A.: Comparative study of different selection techniques in genetic algorithm. Int. J. Eng. Sci. Vo. **6**(3), 174–180 (2017)
27. LedCraft Inc.: How the entertainment industry can use LED displays to create immersive experiences. Official Product Blog (2023). Accessed 02 Dec 2024
28. Lee, K., et al.: Aligning text-to-image models using human feedback. arXiv, abs/2302.12192 (2023)
29. Leonardo AI: Official product page, leonardo.ai. Accessed 02 Dec 2024
30. Leonardo AI: Train a Custom Model, Documentation Page, leonardo.ai. Accessed 02 Dec 2024
31. Lin, T.-Y., Dollár, P., Girshick, R., He, K., Hariharan, B., Belongie, S.: Feature pyramid networks for object detection. In: Proceedings of the 2017 IEEE Conference on Computer Vision and Pattern Recognition (CVPR), Honolulu, pp. 936–944 (2017). https://doi.org/10.1109/CVPR.2017.106
32. Lin, T.-Y., Goyal, P., Girshick, R., He, K., Dollár, P.: Focal loss for dense object detection. In: Proceedings of the 2017 IEEE International Conference on Computer Vision (ICCV), Venice, pp. 2999–3007 (2017). https://doi.org/10.1109/ICCV.2017.324
33. Lisanti, G., Masi, I., Pernici, F., Del Bimbo, A.: Continuous localization and mapping of a pan–tilt–zoom camera for wide area tracking. Mach. Vis. Appl. **27**(7), 1071–1085 (2016). https://doi.org/10.1007/s00138-016-0799-x

34. Liu, W., et al.: SSD: single shot multibox detector. In: Leibe, B., Matas, J., Sebe, N., Welling, M. (eds.) ECCV 2016. LNCS, vol. 9905, pp. 21–37. Springer, Cham (2016). https://doi.org/10.1007/978-3-319-46448-0_2
35. Medjahed, S.A.: A comparative study of feature extraction methods in images classification. Int. J. Image Graph. Signal Process. **7**, 16–23 (2015). https://doi.org/10.5815/ijigsp.2015.03.03
36. Mehrabi, N., et al.: Resolving ambiguities in text-to-image generative models. In: Proceedings of the 61st Annual Meeting of the Association for Computational Linguistics (Volume 1: Long Papers), Toronto, pp. 14367–14388 (2023)
37. Moody, J., Dexter, P.: Concert Lighting: The Art and Business of Entertainment Lighting. 4th edn. Routledge (2016). https://doi.org/10.4324/9781315672816
38. Novelli, C., Casolari, F., Hacker, P., Spedicato, G., Floridi, L.: Generative AI in EU law: liability, privacy, intellectual property, and cybersecurity. Comput. Law Secur. Rev. **55**, 106066 (2024). https://doi.org/10.1016/j.clsr.2024.106066
39. Oliszewski, A., Fine, D., Roth, D.: Digital Media, Projection Design, and Technology for Theatre. 1st edn. Routledge, New York (2018). https://doi.org/10.4324/9781315666976
40. Pires, F., Silva, R., Raposo, R.: A survey on virtual production and the future of compositing technologies (2022). https://api.semanticscholar.org/CorpusID:252571929
41. Prompthero: Openjourney V4, Official Hugging Face Page. Accessed 02 Dec 2024
42. Redmon, J., Farhadi, A.: YOLOv3: an incremental improvement. arXiv, abs/1804.02767 (2018)
43. Redmon, J., Divvala, S., Girshick, R., Farhadi, A.: You only look once: unified, real-time object detection. In: Proceedings of the 2016 IEEE Conference on Computer Vision and Pattern Recognition (CVPR), Las Vegas, pp. 779–788 (2016). https://doi.org/10.1109/CVPR.2016.91
44. Ren, S., He, K., Girshick, R., Sun, J.: Faster R-CNN: towards real-time object detection with region proposal networks. IEEE Trans. Pattern Anal. Mach. Intell. **39**(6), 1137–1149 (2017). https://doi.org/10.1109/TPAMI.2016.2577031
45. Rezatofighi, H., Tsoi, N., Gwak, J., Sadeghian, A., Reid, I., Savarese, S.: Generalized intersection over union: a metric and a loss for bounding box regression. In: Proceedings of the 2019 IEEE/CVF Conference on Computer Vision and Pattern Recognition (CVPR), Long Beach, pp. 658–666. https://doi.org/10.1109/CVPR.2019.00075
46. Sandler, M., Howard, A., Zhu, M., Zhmoginov, A., Chen, L.C.: MobileNetV2: inverted residuals and linear bottlenecks. In: Proceedings of the 2018 IEEE/CVF Conference on Computer Vision and Pattern Recognition, Salt Lake City, pp. 4510–4520 (2018). https://doi.org/10.1109/CVPR.2018.00474
47. Sauer, A., et al.: Adversarial diffusion distillation. In: Leonardis, A., Ricci, E., Roth, S., Russakovsky, O., Sattler, T., Varol, G. (eds.) ECCV 2024. LNCS, vol. 15144, pp. 87–103. Springer, Cham (2024). https://doi.org/10.1007/978-3-031-73016-0_6
48. Shan, S., Ding, W., Passananti, J., Wu, S., Zheng, H., Zhao, B.Y.: Nightshade: prompt-specific poisoning attacks on text-to-image generative models. In: Proceedings of the 2024 IEEE Symposium on Security and Privacy (SP), San Francisco, pp. 807–825 (2024). https://doi.org/10.1109/SP54263.2024.00207
49. Shan, X., Chung, J.: Comparison of the characteristics of green screen and LED wall in virtual production system. Int. J. Adv. Smart Convergence **11**(2), 64–70 (2022)
50. Stability AI: Official product Page, Introducing SDXL Turbo: A Real-Time Text-to-Image Generation Model. Accessed 02 Dec 2024

51. Statista: Zoom Video Communications daily meeting participants worldwide from 2019 to 2020 (in millions). Accessed 02 Dec 2024
52. Tan, M., Pang, R., Le, Q.V.: EfficientDet: scalable and efficient object detection. In: Proceedings of the 2020 IEEE/CVF Conference on Computer Vision and Pattern Recognition (CVPR), Seattle, pp. 10778–10787 (2020). https://doi.org/10.1109/CVPR42600.2020.01079
53. Wang, J., et al.: Review of large vision models and visual prompt engineering. Meta-Radiol. **1**(3), 100047 (2023). https://doi.org/10.1016/j.metrad.2023.100047
54. Willment, N., Swords, J.: What is Virtual Production? An Explainer & Research Agenda. XR Stories (2023). https://xrstories.co.uk/wp-content/uploads/2023/01/What-is-VP-final.pdf. Accessed 02 Dec 2024
55. Yaseen, M.: What is YOLOv8: an in-depth exploration of the internal features of the next-generation object detector. arXiv, abs/2408.15857 (2024)
56. Yin, L., Wang, W., Zhao, J.: Real-time video chroma keying: a parallel approach based on local texture and global colour distribution. IET Image Process. **10**, 638–645 (2016). https://doi.org/10.1049/iet-ipr.2015.0450
57. Zeng, X., Gao, Z., Ye, Y., Zeng, W.: IntentTuner: an interactive framework for integrating human intentions in fine-tuning text-to-image generative models. In: Proceedings of the 2024 CHI Conference on Human Factors in Computing Systems (CHI 2024), Article 182, pp. 1–18. Association for Computing Machinery, New York (2024). https://doi.org/10.1145/3613904.3642165
58. Zhao, W., Shi, M., Yu, X., Zhou, J., Lu, J.: FlowTurbo: towards real-time flow-based image generation with velocity refiner. arXiv, abs/2409.18128 (2024)

ChemFlow: Bridging Computational Tools and Bioinformatics Through LLM-Driven Workflow Automation

Stefan-Bogdan Marcu[1(✉)], Yanlin Mi[1], Mark Tangney[3], and Sabin Tabirca[1,2]

[1] School of Computer Science and Informational Technology, University College Cork, Cork, Ireland
{120227147,120220407,Tabirca}@umail.ucc.ie
[2] Faculty of Mathematics and Informatics, Transilvania University of Braşov, Braşov, Romania
Tabirca@unitbv.ro
[3] Cancer Research@UCC, APC Microbiome Ireland, SynBioCentre, iEd Hub, University College Cork, Cork, Ireland
M.Tangney@ucc.ie

Abstract. Innovations in computational methodologies have significantly transformed the landscape of scientific research, *in silico* experiments have replaced some of the physical experiments. ChemFlow, a novel proof-of-concept platform introduced in this paper, coalesces these advancements by automating the creation of workflows. Designed specifically for the bioinformatics field, ChemFlow leverages Large Language Models and prompt engineering techniques to interpret natural language descriptions and convert them into executable workflows without the need for manual coding. Our contributions are two-fold: first, we introduce an innovative workflow generation and execution platform with the help of large language models, and second, we introduce a novel set of prompt optimisation strategies that improve both the accuracy and efficiency of the generated workflows. ChemFlow enables researchers to focus on domain-specific challenges rather than computational intricacies, making it a pivotal tool for advancing scientific productivity and innovation.

Keywords: Workflow Generation · Prompt Optimisation · Large Language Models · LLM · AI

1 Introduction

In the landscape of contemporary scientific exploration, computational methodologies have profoundly reshaped how research is conducted. The digital revolution, prominently featuring *in silico* experiments, has become a cornerstone of modern scientific inquiry, as computational tools and simulations play an increasingly vital role across diverse disciplines [11]. The synergy between sophisticated

D. Simian and L. F. Stoica (Eds.): MDIS 2024, CCIS 2486, pp. 123–137, 2025.
https://doi.org/10.1007/978-3-031-87386-7_9

software systems is crucial, as scientific progress now frequently relies on the intricate interplay of multiple, advanced technological solutions [11]. This integration presents a set of challenges. The rapid pace at which new state-of-the-art (SOTA) methods are discovered adds layers of complexity and demands substantial time and labour, making the wide adoption of these advanced techniques both challenging and resource-intensive [11].

1.1 Motivation

The creation of ChemFlow is founded on the need to enhance the efficiency of the drug discovery process by alleviating the technical burdens associated with deploying SOTA models. By simplifying the integration and use of these complex systems, ChemFlow enables scientists to pivot their focus back on the domain-specific aspects of their research, those that directly contribute to scientific breakthroughs, rather than becoming intimately familiar with the technicalities of software and system integration. ChemFlow is directed at helping the researcher focus on the scientific problem and accelerate innovation.

1.2 Our Contribution

In this paper, our contribution is two-fold. First, we introduce ChemFLow, a proof of concept, innovative platform that transforms natural language descriptions into executable scientific workflows. This capability leverages prompt engineering techniques to seamlessly interpret natural language inputs and convert them into executable workflows. Chemflow features a novel pipeline specifically developed for protein design, which operates without generating any code, thereby avoiding security risks typically associated with code execution.

Second, we enhance the interaction with Large Language Models through a novel set of prompt optimisation strategies. These enhancements are crucial for ensuring the accuracy and effectiveness of the workflows generated by ChemFlow, facilitating more precise and operationally efficient scientific research. By focusing on these components, ChemFLow represents a significant advancement in the field, enabling researchers to focus more on the scientific inquiry itself and less on the underlying computational mechanics.

2 Background and Related Work

2.1 Workflows

A scientific workflow (Fig. 1) is a detailed, step-by-step specification of a set of tasks or processes designed to accomplish a specific goal. These workflows are typically represented in terms of tasks and their dependencies (depicted in figure Fig. 1 the "Structure based Virtual Screening" component depends on information from both "Databases" and "Binding Site Prediction" components), providing a structured framework for managing complex processes [9,11]. Scientific

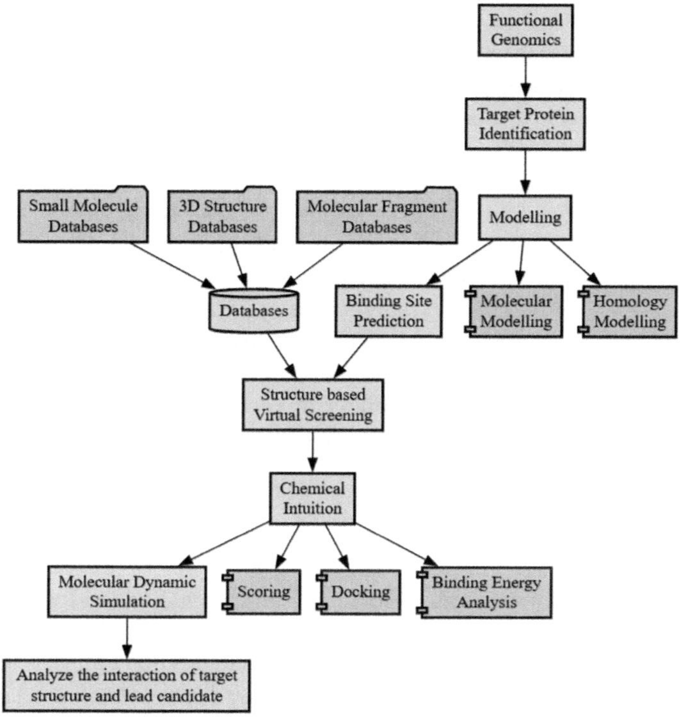

Fig. 1. Workflow of Computation Drug Design. The diagram illustrates an example of a typical computational drug design workflow. This is a demonstrative workflow containing different kinds of operations, modules and dependencies within the typical drug design work.

workflows usually consist of a series of activities involving input, data transformation and output, all directed towards achieving a particular objective. Due to their intrinsic nature, scientific workflows are most commonly represented as directed acyclic graphs [9], which clearly depict the task dependencies and execution order.

The primary purpose of scientific workflows is to accelerate scientific discovery by offering several key advantages, such as automating repetitive tasks, minimising the risk of human error and unshackling the researcher from manually executing monotonous tasks. Another key advantage is their adaptation and reuse property, which enables them to be applied to different experiments or projects, promoting efficiency and consistency in scientific research and providing the researcher with a jumping point in designing the workflow for his specific task.

Scientific workflows can vary in complexity, ranging from simple, linear chains of tasks to intricate, graph-structured dependencies. Regardless of their complexity, all workflows consist of three fundamental components: inputs, transformations, and outputs. The input elements represent the points at which external data is ingested into the process. The transformation elements are the series of computational steps through which the input data is processed and transformed. The output elements are represented by the products, or a subset of their characteristics, generated by the workflow after the dependency chain for the specific output node has been processed.

Workflows lie at the foundation of scientific research, serving as the backbone for systematically conducting experiments and analyses. Meticulously designed experiment protocols ensure that each step is clearly, rigorously defined and followed. This is essential for achieving accurate and reproducible results. Without well-defined workflows, the scientific method would be compromised by inconsistencies and errors, undermining the validity of research findings. By bringing structure and clarity, workflows also help researchers stay organized, facilitating the flawless execution of research activities, enhancing productivity and improving the quality of research.

2.2 Large Language Models

Throughout history, humans have continuously strived to bridge the communication gap with machines. Starting from the abstraction of electrical voltages in wires into binary 1 s and 0 s, forming the basis of machine language. Incrementally creating more complex and building up layers of abstraction evolving assembly instructions that allow a slightly more intuitive communication channel. As the abstraction layers grow and technology advances, we reach high-level programming languages like Python, characterized by their strictly defined syntax and relative ease of use, bringing the communication point between humans and machines even closer to natural language. Large Language Models (LLMs) have the potential to represent the latest milestone in this evolution.

Large Language Models (LLMs), also referred to as foundational models, represent a significant leap in artificial intelligence technologies. These models are primarily built on the transformer architecture, a type of deep learning model that relies on mechanisms like attention [19] to process and model complex data dependencies. LLMs are trained on expansive, internet-level datasets. This training enables them to perform a wide array of natural language processing tasks with remarkable proficiency. The global landscape of artificial intelligence has been profoundly transformed by the models such as ChatGPT [4,16], LLaMA [18] and PaLM [8]. These models excel in generating coherent and contextually appropriate text and demonstrate capabilities in more complex tasks like translation, content creation, and even certain types of reasoning and problem-solving [5], including text generation, summarization and question answering.

The ability of LLMs to understand and generate natural language enables more intuitive interaction with machines, potentially transforming the way researchers design and execute workflows. However, the efficacy of these interactions is highly dependent on the quality of the initial prompt [20,23]. As LLMs continue to integrate into the scientific process, laying the foundation for a future where the symbiotic relationship between machines and researchers is significantly deepened, they drive forward scientific discovery and innovation.

2.3 Related Works

While no existing literature directly addresses the generation of scientific workflows from natural language, several studies provide insights tangential to the topic.

Agrawal et al. and Wei et al. demonstrate LLM's zero-shot, one-shot, and few-shot learning capabilities, highlighting their adaptability to complex tasks with minimal input [1,21]. Carta et al. use LLMs to create knowledge graphs from unstructured data, showcasing their ability to synthesize and structure information [6]. These studies underline LLMs' potential in generating workflows from natural language without requiring any expensive model training.

Zeng et al. introduce FlowMind, an LLM-based system for Robotic Process Automation, which aims to automate repetitive tasks using context-grounding prompts [24]. With a concept superficially similar to ChemFlow, the two approaches differ significantly in the usage of the LLM and in the workflow execution approach. ChemFlow applies novel prompt engineering without generating code, mitigating the potential security risks associated with it, and primarily targeting the field of bioinformatics.

In the realm of code generation, models like GitGub Copilot and CodeGen demonstrate LLMs' capabilities to comprehend API description protocols and generate complex structures [15,25]. Bubeck et al. highlight GPT4's broad abilities, emphasizing its potential in reasoning and complex task execution [5]. These examples illustrate the feasibility of using LLMs for structured output generation.

Prompt engineering, as discussed by Wang et al., Sahoo et al., and Chen et al., provides methodologies to optimize LLM interactions and enhance its performance in specialized tasks [7,17,20].

ChemFlow leverages the use of LLMs in the use of natural language for workflow generation, employing concepts from fields such as LLM adaptability, information structuring, and prompt engineering in order to inform its development and underscore its innovative approach.

3 Proposed Method

We propose ChemFlow, a platform characterized by a modular architecture segregated into three functional components: the frontend, the backend, and integration with a Large Language Model provider, OpenAI GPT through an API interface. The architecture of ChemFlow focuses on interoperability and ease of extension through a plugin interface that employs LLMs in an ideal format, leveraging their automation power, translational properties and knowledge while negating issues like hallucinations [10] through the use of grounding, automated validation and user-provided correction and validation.

Generating workflows with the help of ChemFlow offers a significant enhancement over traditional scripting approaches. One of the advantages of using this platform is that it meticulously captures both the computational graph and the parameters used in the workflow, ensuring a high degree of reproducibility and traceability. The platform also provides uniform access to computational resources, avoiding the inconsistencies commonly associated with ad hoc scripting methods. Additionally, ChemFlow simplifies the complexities associated with managing multiple compute nodes, various communication protocols, and different job scheduling systems.

Another significant advantage brought forward by the flexible integration of the LLM provider together with the API call-based workflow design is the option to deploy the system locally. This represents a great security feature as during the conception of the workflow and during its execution, there is no data that leaves the deployment network. This property is of significant importance in any case that deals with sensitive data.

3.1 Frontend

The frontend of ChemFlow is developed using the Vue.js framework and the Vuetify component library, offering a streamlined and user-friendly interface (depicted in Fig. 2). The interface supports three methods for workflow creation: generating workflows from natural language descriptions using the LLM, utilizing the intuitive drag-and-drop interface inspired by the visual programming paradigm, and importing and modifying existing workflows to facilitate iterative and collaborative work. The workflow illustrated in Fig. 2 is the result of two of the three modes of interaction; the initial graph was generated from the "Starting from the '1a3n' protein and from other proteins with a GRAVY score of 0.03 and a monoisotopic mass of 15 Da generate new sequences with an Instability Index under 15; fold the sequences and send the results by email to test@email.com" description resulting in the displayed workflow without the edge between the PS-GO component and the Fold component, which has been manually added through the drag and drop functionality. The interface is designed with the goal of simplifying the initial design process while providing visual representations for detailed inspection and necessary adjustments.

A key feature of the frontend is represented by its drag-and-drop functionality, which allows users to create or edit existing workflows by adding or removing nodes or connections of the existing workflow. Figure Fig. 2 illustrates a snapshot of the User Interface with a relatively simple workflow depicted. The workflow shown exemplifies a process that employs the currently implemented nodes, searching the RCSB PDB [2] database based on its ID, performing parametric searches with PS-GO [14], fine-tuning parameters using the Silver Surfer [12], generating 3D protein structures with the help of PROFASA [13] and emailing the results. The available nodes are manually developed to demonstrate the feasibility of the project and to explore the experience and functional requirements. Future work will explore the challenge of employing LLMs to automate

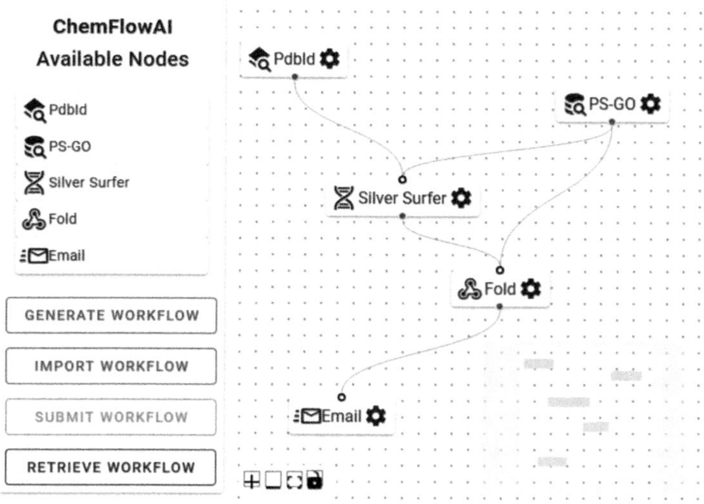

Fig. 2. Screenshot of the Workflow Design GUI of ChemFLow. The snapshot presents the workflow design view of the ChemFlow frontend with a simple demonstrative workflow containing the PdbID, PS-GO, SilverSurfer, Fold and Email components and their defined dependencies. On the left side of the panel are presented the different nodes that can be used within the design, supporting a drag and drop interface, together with buttons for generating a workflow from a natural language description, for importing a previously saved workflow, for submitting the workflow for execution, and for inquiring about the state of a workflow that has been previously submitted for execution.

the deployment of servers given a code repository and a containerisation interface. The web-based dashboard offers real-time visualization of workflow states (depicted in Fig. 3), ensuring users can monitor and manage their workflows effectively.

3.2 Backend

The backend of the ChemFlow platform is developed using the Python language and utilizing a RESTful API via Flask and a REDDIS in memory store. It supports robust integration of LLMs accessed through an API. The backend is designed around API calls to different servers that provide the functionality for each of the available nodes, with parameters provided by the LLM and the data bus system that manages both incoming and outgoing data, maintaining the continuity and integrity of information throughout the workflow execution process. In order to keep sensitive data within organizational boundaries and enhance the security posture of the platform, it supports a dual deployment strategy, with options for both in-the-cloud and on-premises installation.

The backend provides two major functionalities of the platform: an orchestrator in charge of workflow management and execution, and a natural-language-based workflow designer. The orchestrator handles the entire workflow execution

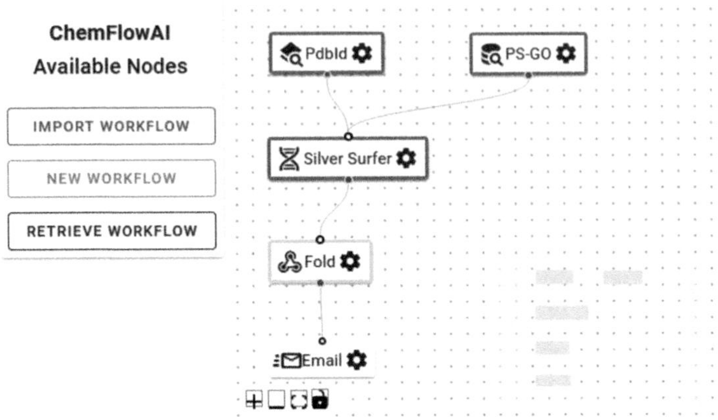

Fig. 3. Screenshot of the Workflow Visualisation GUI of ChemFlow. The image depicts the workflow status view of the ChemFlow frontend with a simple workflow in the process of being executed. Depicted are 3 processes that have finished processing, PdbId, PS-GO and Silver Surfer, denoted with a green border; 1 process running, Fold, marked with a yellow border; and 1 Process, E-mail, waiting for all of its dependencies to finish processing. (Color figure online)

lifecycle. Once a workflow has been submitted for execution, the orchestrator performs a dependency sort and starts a parallel execution for each of the nodes that have no dependencies. When the orchestrator is notified by a node of the processing being finalised, it updates the node's status, places the generated data on the data bus, and updates all the dependencies. Subsequently, all nodes with resolved dependencies are executed in parallel until all of the nodes of the workflow have been exhausted. The status of each node and the generated data are synchronised and managed through the help of the REDDIS store.

For workflow design, the backend employs the LLM at two different abstraction levels: Graph Generation and Node Configuration. During Graph Generation, the LLM produces a computational graph based on the functionalities of the available plugins and the description of the workflow provided by the user. The graph is iteratively computationally validated and optimised. For Node Configuration, the workflow description and the graph are provided to the network in order to generate the appropriate specific configurations for each of the nodes within the graph. Even though multiple static checks are performed in order to validate the syntactic correctness of the generated workflow, the possibility of syntactically correct but nonsensical results is not eliminated. Therefore, the graphical representation of the workflow presented by the frontend serves as a final guard against LLM hallucinations; the workflow is presented to the user for validation and editing before submission for execution. In order to extract as much functionality as possible out of the language model, a series of foundational model optimisation strategies have been employed.

3.3 LLM Optimisation Strategies

In the development of ChemFlow, optimizing the performance of the LLM is crucial for ensuring accurate workflow generation. To maximize the model's one-shot performance, we have devised a novel combination of prompt strategies. These strategies include Prompt Style Transfer, Prompt Rephrasing, Prompt Iterative Improvement, Prompt Ranking and Prompt Correcting (Fig. 4).

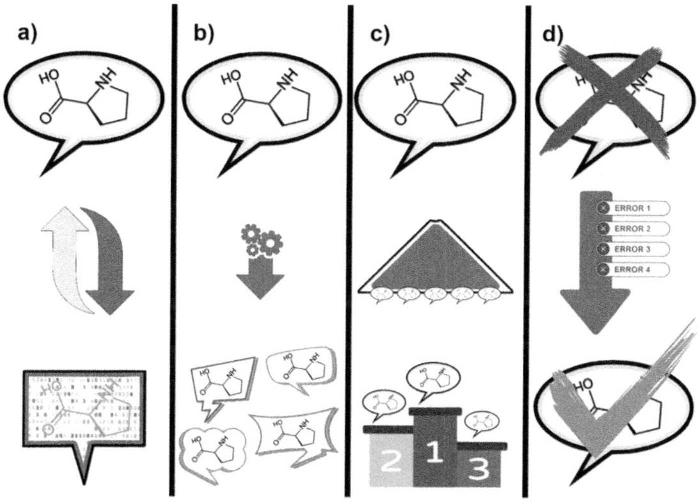

Fig. 4. The Figure conceptually illustrates the four prompt optimisation strategies employed by the ChemFlow platform. a represents Prompt Style Transfer, which reformulates the task using a different lingo; b depicts Prompt Rephrasing, the genetic algorithm-inspired approach of slightly rephrasing the task; c presents the Prompt Iterative Improvement and Ranking strategy, in which the network is tasked to propose improvements to an existing solution and then rank the solutions based on the fidelity in respect to the initial description and the optimal use of resources; d displays the strategy of Prompt Correcting, in which a wrong solution is provided together with the errors and the LLM is tasked with producing a valid solution.

Prompt Style Transfer is a strategy employed within ChemFlow which involves directing the LLM to reinterpret and express the initial workflow description using the specialized terminology and style intrinsic to computer science. This strategy is partly inspired by the observation that LLMs exhibit variable performance across different languages, with marked proficiency in English [23]. Expanding on this, we propose that using expert technical language substantially enhances the quality of the generated results.

Our empirical observations reinforce the idea that technical language significantly improves the output generated by the LLM. This improvement might be born from the fact that an LLM operates as an ensemble of roles, where the usage of precise technical terminology effectively signals the model to assume a more 'technical' role. This phenomenon is supported by the findings of Bhargava et al. in their "What's the Magic Word? A Control Theory of LLM Prompting" study [3], and is further corroborated by Yin et al., who note that the effectiveness of LLMs is influenced not just by the language used but also by the level politeness in the language.

We hypothesize that this effect stems from the training data biases and the context-switching capabilities of LLMs; by using the correct technical language, the model is encouraged to assume the appropriate role for the task. This novel method of prompt engineering improves the relevance and accuracy of the generated workflows.

Prompt Rephrasing is a novel technique utilised within ChemFlow that enhances the utility of the LLM by generating multiple alternative descriptions that retain the fundamental description of the initial input. This method was inspired by the observation that minor modifications, particularly at the beginning of a prompt, can significantly alter both the performance and the directional focus of the LLM's responses. This observation aligns with the findings presented by Bhargava et al. [3], who highlighted the impact of the initial prompt modifications on the output of the LLM.

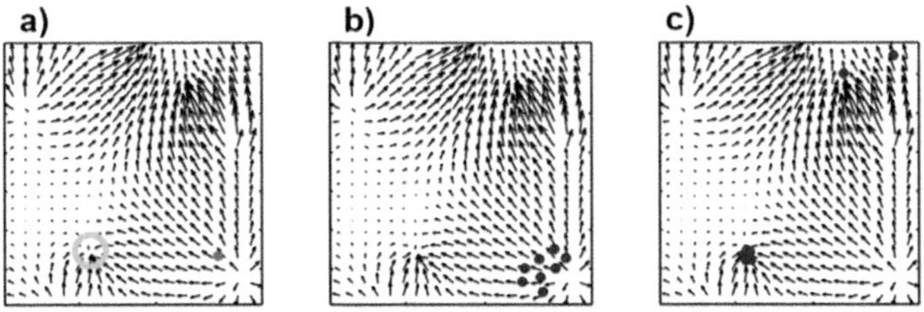

Fig. 5. Visual description of the prompt rephrasing strategy inspiration. The task of producing an optimal response from the LLM is envisioned as navigating a flow field towards an area containing correct answers. a depicts a conceptual low-dimension representation of the network's inherent internal flow field with a green dot marking the spot arrived at from the processing of the initial prompt, and the gold circle the area where correct answers can be found; b depicts with blue marks the spots arrived at after the processing of the slightly rephrased initial prompts; c demonstrates that after allowing the LLM to produce the answer, a process akin to following the path of the flow field, multiple of the answers arrive in the area depicting correct answers. (Color figure online)

The underlying concept of Prompt Rephrasing borrows principles from Genetic Algorithms that require an initial population of different individuals to speed up the convergence towards a minimum, and where a good spread of the initial population helps the algorithm surpass local minima. By envisioning the generation of the next token by the LLM as akin to adding a vector in a direction vaguely defined by a semi-ordered high-dimensional flow field defined by the model's architecture (depicted in figure Fig. 5 subpoint a, with the gold circle delimiting the hypothetical area of good answer and the green dot representing the initial prompt), the rationale for Prompt Rephrasing becomes evident. Starting with multiple similar but slightly varied descriptions of the same concept (depicted in figure Fig. 5 subpoint b as blue dots) essentially provides several initial vectors within a closely related area within the high-dimensional space. This increases the likelihood that at least one of the starting points will navigate the flow field towards a more accurate and relevant output (illustrated in figure Fig. 5 subpoint C).

This technique leverages the diversity of expression to explore a broader range of potential solutions, thus increasing the chances of arriving at an optimal response. By implementing Prompt Rephrasing, ChemFlow strives to more robustly generate workflows that are optimal and align with the user's workflow specification.

Prompt Iterative Improvement and Ranking is a method leveraged by ChemFlow that revolves around the iterative refinement of workflow solutions. This technique takes advantage of the model's ability to optimise pre-existing solutions. Specifically, it begins by providing the LLM with an initial workflow solution and taking the LLM to suggest new solutions that represent a more optimized version of the initial input.

In conjunction with Iterative improvement, Ranking plays a pivotal role in evaluating the quality of generated workflows. This aspect of the strategy involves presenting the LLM with multiple potential solutions and requesting that it rank these alternatives based on predefined criteria. In the context of ChemFlow, these conditions focus on the fidelity to the initial prompt and their efficiency.

Inspiration for this strategy is drawn from the conceptual experiments described in the "Sparks of artificial general intelligence: Early experiments with GPT-4" work of Bubeck et al. [5], which explores the capacity of GPT-4 to exhibit some aspects of general intelligence and the works in prompt engineering of Yao et al. [22]. By integrating this approach, ChemFlow harnesses the LLM's self-improvement capabilities, grounding it with an existing example. This approach enhances the reliability, precision and optimisation of the produced workflows.

Prompt Correcting is an optimisation used within the ChemFlow platform that enhances the accuracy of the LLM by prompting the model to correct errors. This technique involves presenting the LLM with a previously generated incorrect response along with the associated error message generated either by an API call or through computational validation. The task for the LLM is then to rectify the inaccuracies by revising the initial output.

We hypothesize that the effectiveness of Prompt Correcting can be attributed to two main factors. Firstly, the training corpus of the LLMs includes a substantial amount of forum conversational data, such as from Stack Overflow. This type of data is rich in problem-solving interactions, where errors are commonly identified and corrected within the dialogue. By structuring the prompt to mimic these scenarios, the model is encouraged to adopt a problem-solving role.

Secondly, the focused nature of the error itself is crucial. By providing the model LLM with the context of where and how the error occurred, encapsulated within the error message, the model receives critical insights into the nature of the mistakes and empowers the model to generate a valid solution.

Through the application of the Prompt Correcting strategy, ChemFlow grounds the LLM and minimises the apparitions and impact of hallucinations [10].

4 Results

During the development and evaluation of ChemFlow, a comprehensive assessment was conducted through internal testing, feedback from users, and small-scale experimental tests. Users interacted with ChemFlow to design and execute workflows utilizing five implemented nodes: PDBID Search, PS-GO [14], Silver Surfer [12] parametric engineering, ProteinFolding provided by PROFASA [13] and e-mail presentation. Their feedback provided insights into the platform's user experience and effectiveness.

4.1 Frontend

The platform's drag-and-drop interface was highlighted for its simplicity and intuitiveness, allowing users to construct and modify workflows easily. Participants appreciated the ability to visually map out their workflows, which aided in understanding and managing the processes more effectively. Additionally, the real-time visualization feature was well-received, offering users valuable feedback on the execution status of their workflow.

4.2 Backend

The robustness of ChemFlow's backend was a critical area of focus during the development phase. The backend's reliable orchestrator was crucial in efficiently managing workflow execution, demonstrating a high capacity for handling diverse workloads ranging from simple to complex scenarios. The system does not have a high computational footprint; its role focuses on delegating tasks via API calls and collecting and centralising the generated information. The dual deployment option, allowing for both cloud-based and on-premise installations, was particularly appreciated, offering flexibility and enhanced security for handling sensitive data.

4.3 Workflow Generation

While large-scale tests were not conducted due to their high costs and challenges in generating a realistic and accurate dataset, small-scale tests were instrumental in evaluating the workflow generation capabilities of ChemFlow. We have manually generated 40 description-graph pairs for testing that have been augmented through LLM rephrasing and manual validation to 400 test cases. In these tests, the generated workflows were found to be correct and optimal in 90% of the cases. 8% of the generated workflows were correct but included unnecessary steps, and the remaining 2% constituted cases where the workflow generation failed to produce the required information.

The empirical assessment revealed that users highly valued the ability to generate workflows from natural language descriptions, noting that the translation from natural language to workflows performed surprisingly well. This feature was seen as a significant advantage, likely to be utilized by users in their day-to-day work for its convenience and effectiveness in generating workflows.

5 Conclusion and Future Works

ChemFlow stands as evidence of the progress in utilizing LLMs to facilitate direct and efficient communication between humans and machines, particularly in automating and managing scientific workflows without requiring advanced IT knowledge, thus allowing the researcher to focus on his area of expertise. This platform enhances the accuracy and speed of scientific discovery by allowing intuitive workflow management and reducing the potential for human error. Through its robust architecture, ChemFlow delivers a reliable system capable of managing both simple and complex tasks, where its user-friendly front end enables the easy construction and management of workflows.

Looking ahead, the development of ChemFlow will focus on several directions in order to broaden its utility and efficiency. We plan to further facilitate the reuse of workflows by developing an integrated library of bioinformatic workflows that will serve as ready-to-use templates for common tasks. Another direction of developing this proof of concept is by expanding the functionality set through the addition of more servers that implement diverse operations of general use in the processing of proteins. We also aim to experiment with eager execution, allowing the system to start processing nodes using partial data available; this has the potential to further optimize the execution time and resource utilization. A different direction we are excited to explore is the use of LLMs for automatically configuring and deploying Docker instances to expand even more the functionality and applicability of the ChemFlow platform. These directions provide exciting prospects for the future of ChemFlow.

Acknowledgments. This study was funded by the University College Cork, School Of Computer Science.

Disclosure of Interests. The authors have no competing interests to declare that are relevant to the content of this article.

References

1. Agrawal, M., Hegselmann, S., Lang, H., Kim, Y., Sontag, D.: Large language models are few-shot clinical information extractors. In: Goldberg, Y., Kozareva, Z., Zhang, Y. (eds.) Proceedings of the 2022 Conference on Empirical Methods in Natural Language Processing, pp. 1998–2022. Association for Computational Linguistics, Abu Dhabi (2022). https://doi.org/10.18653/v1/2022.emnlp-main.130
2. Berman, H.M., et al.: The protein data bank. Nucleic Acids Res. **28**(1), 235–242 (2000). https://doi.org/10.1093/nar/28.1.235
3. Bhargava, A., Witkowski, C., Shah, M., Thomson, M.: What's the magic word? A control theory of LLM prompting. arXiv preprint arXiv:2310.04444 (2023). https://doi.org/10.48550/arXiv.2310.04444
4. Brown, T.B., et al.: Language models are few-shot learners (2020)
5. Bubeck, S., et al.: Sparks of artificial general intelligence: early experiments with GPT-4 (2023). https://doi.org/10.48550/arXiv.2303.12712. https://www.microsoft.com/en-us/research/publication/sparks-of-artificial-general-intelligence-early-experiments-with-gpt-4/
6. Carta, S., Giuliani, A., Piano, L., Podda, A.S., Pompianu, L., Tiddia, S.G.: Iterative zero-shot LLM prompting for knowledge graph construction (2023). https://doi.org/10.48550/arXiv.2307.01128
7. Chen, B., Zhang, Z., Langrené, N., Zhu, S.: Unleashing the potential of prompt engineering in large language models: a comprehensive review (2023). https://doi.org/10.48550/arXiv.2310.14735
8. Chowdhery, A., et al.: Palm: scaling language modeling with pathways (2022)
9. Deelman, E., Gannon, D., Shields, M., Taylor, I.: Workflows and e-science: an overview of workflow system features and capabilities. Future Gener. Comput. Syst. **25**(5), 528–540 (2009). https://doi.org/10.1016/j.future.2008.06.012. https://www.sciencedirect.com/science/article/pii/S0167739X08000861
10. Ji, Z., et al.: Survey of hallucination in natural language generation. ACM Comput. Surv. **55**(12) (2023). https://doi.org/10.1145/3571730
11. Ludäscher, B., Weske, M., McPhillips, T., Bowers, S.: Scientific workflows: business as usual? In: Dayal, U., Eder, J., Koehler, J., Reijers, H.A. (eds.) Business Process Management, pp. 31–47. Springer, Heidelberg (2009)
12. Marcu, S.B., Mi, Y., Yallapragada, V.V.B., Tangney, M., Tabirca, S.: Silver surfer: navigating the parametric protein space with genetic algorithms. In: The 37th International Conference on Industrial, Engineering & Other Applications of Applied Intelligent Systems (2024)
13. Mi, Y., Marcu, S.B., Tabirca, S., Yallapragada, V.: Profasa-a web-based protein fragment and structure analysis workstation. Front. Bioeng. Biotechnol. **11**, 1192094 (2023). https://doi.org/10.3389/fbioe.2023.1192094
14. Mi, Y., Marcu, S.B., Tabirca, S., Yallapragada, V.V.: PS-GO parametric protein search engine. Comput. Struct. Biotechnol. J. **23**, 1499–1509 (2024). https://doi.org/10.1016/j.csbj.2024.04.003. https://www.sciencedirect.com/science/article/pii/S2001037024000886
15. Nijkamp, E., et al.: Codegen: an open large language model for code with multi-turn program synthesis (2023). https://doi.org/10.48550/arXiv.2203.13474
16. OpenAI, et al.: GPT-4 technical report (2024)
17. Sahoo, P., Singh, A.K., Saha, S., Jain, V., Mondal, S., Chadha, A.: A systematic survey of prompt engineering in large language models: Tech. Appl. (2024). https://doi.org/10.48550/arXiv.2402.07927

18. Touvron, H., et al.: Llama: open and efficient foundation language models (2023)
19. Vaswani, A., et al.: Attention is all you need. In: Proceedings of the 31st International Conference on Neural Information Processing Systems, NIPS 2017, pp. 6000–6010. Curran Associates Inc., Red Hook (2017)
20. Wang, L., et al.: Investigating the impact of prompt engineering on the performance of large language models for standardizing obstetric diagnosis text: Comparative study. JMIR Form. Res. **8**, e53216 (2024). https://doi.org/10.2196/53216
21. Wei, X., et al.: Zero-shot information extraction via chatting with ChatGPT (2023). https://doi.org/10.48550/arXiv.2302.10205
22. Yao, S., et al.: Tree of thoughts: deliberate problem solving with large language models. arXiv abs/2305.10601 (2023). https://doi.org/10.48550/arXiv.2305.10601. https://api.semanticscholar.org/CorpusID:258762525
23. Yin, Z., Wang, H., Horio, K., Kawahara, D., Sekine, S.: Should we respect LLMS? A cross-lingual study on the influence of prompt politeness on LLM performance (2024). https://doi.org/10.48550/arXiv.2402.14531
24. Zeng, Z., et al.: Flowmind: automatic workflow generation with LLMS. In: Proceedings of the Fourth ACM International Conference on AI in Finance, ICAIF 2023, pp. 73– 81. Association for Computing Machinery, New York (2023). https://doi.org/10.1145/3604237.3626908
25. Zhao, S.: Github copilot chat beta now available for all individuals — github.blog. https://github.blog/2023-09-20-github-copilot-chat-beta-now-available-for-all-individuals/. Accessed 21 May 2024

Augmentation and Substitution of Medical Training Data with Generative Adversarial Networks for Machine Learning

Branislav Radomirovic, Luka Jovanovic, Nebojsa Budimirovic, Miodrag Zivkovic, Nebojsa Bacanin(✉), and Milos Dobrojevic

Faculty of Informatics and Computing, Singidunum University, Belgrade, Serbia
{branislav.radomirovic.11,nebojsa.budimirovic.20}@singimail.rs,
{ljovanovic,mzivkovic,nbacanin,mdobrojevic}@singidunum.ac.rs

Abstract. This research underscores the pivotal role of AI in addressing intricate problems, especially in medical diagnosis and predicting treatment outcomes. Despite challenges in algorithm selection, hyperparameter tuning, and limited medical data, simulations with real-world data affirm the effectiveness of machine learning, notably showcasing the random forest model with an impressive 85.19% accuracy. A significant aspect of this work involves the exploration of Generative Adversarial Networks (GANs) for data augmentation and synthesis. GANs enhance classifiers like MLP and AdaBoost but present challenges for decision tree and KNeighbors models. Additionally, leveraging fully synthetic data for training proves promising, offering a potential solution to data scarcity. Feature importance analysis emphasizes the impact of treatment frequency on patient outcomes, enhancing model interpretability. In conclusion, this research addresses challenges, introduces novel GAN-based approaches, and provides valuable insights to advance practical AI applications, particularly in effective data augmentation and synthesis using GANs, ultimately improving the prediction of treatment outcomes.

Keywords: Machine learning · Generative Adversarial Networks · Feature importance · Medical data · Data augmentation

1 Introduction

Artificial intelligence (AI) has, in recent years, permeated several spheres of research. Increased accessibility to computational resources and the availability of techniques have led to a surge in research. The application of AI has benefited multiple fields where complex and non-linear problems need to be addressed [25]. An additional advantage is that AI can learn from data observation with little to no explicit programming. An interesting field of applied AI is in medical diagnosis. By observing expert-labeled data, AI can consistently make decisions that can aid professionals in the field. Another interesting application of AI is predicting treatment outcomes [1].

D. Simian and L. F. Stoica (Eds.): MDIS 2024, CCIS 2486, pp. 138–152, 2025.
https://doi.org/10.1007/978-3-031-87386-7_10

However, there are several challenges that need to be addressed for AI to be successfully applied. Determining an appropriate algorithm, hyperparameter selection, and data availability are just some of the challenges faced by researchers. While the former problems have several techniques that can be applied to mitigate the challenge, such as metaheuristics to select proper parameters or taking a trial-and-error approach [18], the latter problem is unique. Data availability, especially medical data availability, is a foremost challenge in the field. Patient privacy and the cost of medical research limit the amount of high-quality data available to researchers [23]. Furthermore, certain conditions that are particularly rare often lack appropriate sample sizes to facilitate proper algorithm training and evaluation.

This work explores the utilization of emerging generative adversarial networks for their ability to generate high-quality synthetic data. The ability of these networks is explored for medical data augmentation as well as data substitution. In cases where an insufficient number of samples is available, such as rare medical conditions or privacy concerns limiting availability, generative networks can be applied to generate viable training data to facilitate algorithm training. Furthermore, in cases where data cannot be made available, a fully synthetic dataset might be made available to further researchers and the evaluation of viable solutions with little to no cost and privacy concerns. Synthetic data generation has been previously widely utilized in different application domains [3,20,22,31,33].

In this work, a publicly available medical cryotherapy treatment dataset is utilized to test the viability of GANs for data augmentation and substitution. This dataset is particularly interesting as it is fairly small, with only about 90 samples, limiting the number of training samples. Hence, augmentation can help improve algorithm performance [8]. Additionally, data substitution is explored. Three experiments are carried out. In the first experiment, a battery of algorithms is trained on real-world data and tested on real samples using standard metrics. The following two experiments are conducted using augmented data for model training as well as fully synthetic data. In both cases, the models trained on generated data are evaluated on real samples, and their performance is assessed under identical conditions. The contributions of this work may be summarized as follows:

- The exploration of machine learning algorithms of cryotherapy treatment outcome prediction
- The application of GAN for medical data augmentation
- The use of GAN for medical full data substitution

The remainder of this work has the following structure: in Sect. 2, preceding works that helped make this work a reality are explored. In Sect. 3, the methodology explored in this work is presented. Section 4 describes the conducted experiments, and Sect. 5 concludes this work and presents proposals for future research.

2 Related Works

The ability of AI algorithms to learn through data observations has been leveraged in several fields in recent years. Interesting applications for addressing complex issues have been proposed in cybersecurity [2,6,15]. The ability of AI to adapt to a changing environment makes these algorithms suitable for a dynamic landscape, such as cybersecurity. The adaptability of algorithms has made them popular in the financial sector as well, where models have been used to predict prices and potential credit card churn [16]. Nevertheless, a very important application of AI is in medicine, where algorithms are used to aid medical professionals in improving early detection and diagnosis of medical conditions in patients, thereby enhancing patient outcomes.

An additional challenge for algorithm use is the high reliance of AI on proper adjustments. Usually, algorithms are developed with a high degree of flexibility to allow them to be applied to a diverse range of problems. However, generic use yields less favorable outcomes. Therefore, algorithms are published with a set of hyperparameters that can be adjusted to make the algorithms suit the given problem. The challenge of tuning these hyperparameters can be considered NP-hard and therefore not directly solvable through discrete methods. Metaheuristics have proven an interesting approach for addressing this issue in reasonable time frames and realistic periods of time [7,14].

An increasingly popular form of AI is generative algorithms. Using large datasets, models are able to approximate and create new samples that resemble existing instances of data. This has been best exemplified by image generation models such as stable diffusion [28] used to generate artwork, as well as large language models used to generate text. However, the use of generative models has yet to be explored in other areas, such as the application of GAN for sensitive data substitution, as well as data augmentation in cases where samples are scarce or unavailable.

While widely employed in image and text generation models, the exploration of generative models in diverse domains, including the application of GANs for sensitive data substitution and data augmentation in scenarios with limited or unavailable samples such as medical datasets, remains a literature gap.

3 Methods

3.1 Generative Adversarial Networks

A family of deep learning models known as Generative Adversarial Networks [5,9,10] is made up of two neural networks, the discriminator D and the generator G, which are trained concurrently using adversarial training. In order to produce data that are indistinguishable from real data, the discriminator network learns to distinguish between real data samples $\mathbf{x}$ and synthetic data samples $\hat{\mathbf{x}}$. The generator network learns to generate synthetic data samples $\hat{\mathbf{x}} = G(\mathbf{z})$ from random noise $\mathbf{z}$. The generator's formal goal is to minimize the loss function that follows:

$$\min_{G} \max_{D} V(D, G) = \mathbb{E}_{\mathbf{x} \sim p_{\text{data}}(\mathbf{x})}[\log D(\mathbf{x})] + \mathbb{E}_{\mathbf{z} \sim p_{\mathbf{z}}(\mathbf{z})}[\log(1 - D(G(\mathbf{z})))] \quad (1)$$

where the distribution of actual data samples is indicated by $p_{\text{data}}(\mathbf{x})$ and the distribution of the input noise vector, $\mathbf{z}$, is shown by $p_{\mathbf{z}}(\mathbf{z})$. The discriminator accurately classifies synthetic and real samples in an attempt to optimize this goal.

The generator and discriminator play a minimax game where the discriminator improves its ability to distinguish between genuine and fake samples while the generator tries to produce increasingly realistic samples to trick the discriminator. This is done through iterative optimization. Converging to a Nash equilibrium, when the generator generates data that are statistically comparable to actual data and the discriminator is unable to consistently discriminate between real and synthetic samples, is the ultimate goal of GAN training. When the generator produces samples that correspond to the data distribution of actual data, that is, when $G(\mathbf{z})$ resembles the distribution $p_{\text{data}}(\mathbf{x})$, this equilibrium is reached.

The generator and discriminator are in continual communication with each other during the training process in the following ways:

- Synthetic data is created by the generator and sent to the discriminator.
- The discriminator attempts to accurately categorize data by evaluating both actual and bogus information.
- The generator enhances its capacity to provide more realistic data by utilizing the discriminator's feedback. The item This cycle keeps going until the discriminator is unable to consistently separate the data that the generator generates from real data that is almost identical to it.

Once a new sample is generated by G and the D deems it to be a viable sample a classifier (C) is empled to assign an appropriate label to said sample

This process results in a generator that can produce realistic data samples. However, training GANs can be challenging, and achieving stable convergence is an ongoing area of research. The training procedure can be summarized as shown in Algorithm 1.

Algorithm 1. Pseudocode for GAN Training

```
Initialize generator G
Initialize discriminator D
Initialize classifier C
Train C using real data
while Loss values do not converge do
    Determine real samples X_sample
    Initialized batch input G, Z_sample
    Assess inputs based on D decisions
    Apply labels based on C
    Calculate loss function
    Refresh G and D
end while
```

4 Experimental Setup and Results

To facilitate experimentation and validate the introduced approach, a publicly available dataset [19] is utilized[1]. The dataset comprises cryotherapy patient data, including information on patient gender, age, instances of treatment, wart count, type, and affected area. Additionally, the result of the treatment is also available and serves as the target for classification, where value 1 indicates successful treatment and value 0 corresponds to failed treatment. The dataset is fairly small consisting of 90 samples. However the data is reasonably well balanced as shown in Fig. 1.

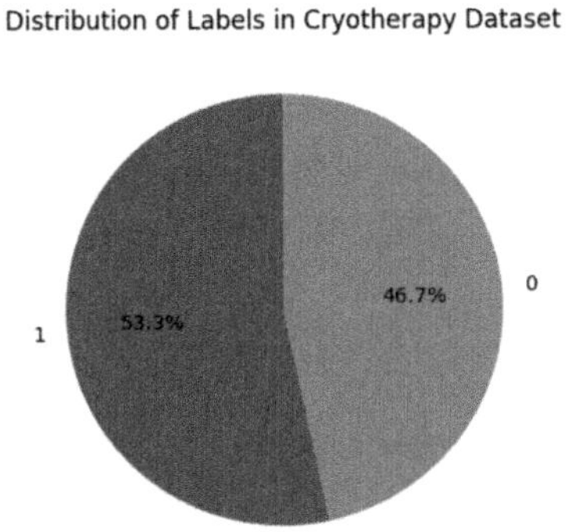

Fig. 1. Dataset balance.

Several machine learning classifiers are employed to predict treatment outcomes, including AdaBoost [11], Decision Trees [4], Gradient Boosting [24], KNeighbors [27], and Random Forest [30]. Additionally, the Support Vector Classifier (SVC) [12] and Multilayer Perceptron (MLP) [29] are evaluated. The algorithms are implemented in Python, utilizing the suggested parameters provided by scikit-learn. Simulations are conducted under identical conditions, with 70% of the data used for testing and reserving 30% for evaluation. The GAN used in this network is implemented using a publicly available TableGan library [26] based on deep convolutional neural networks. Models are implemented using default parameter selections proposed in the work that introduced the approach. The simulations were executed on the configuration based on Intel i9-13900K processor and 128 GB of RAM memory.

A total of three experiments are carried out. The first experiment evaluates each classifier using real-world data to train models and test their performance. The second experiment uses synthetic data to expand the training samples, enhancing the models' overall performance. Finally, in the third simulation, fully synthetic data is employed to train models that are then evaluated on real-world data to test the feasibility of complete data substitution. Models

[1] https://archive.ics.uci.edu/dataset/429/cryotherapy+dataset.

are evaluated using a standard set of evaluation metrics used for classification including accuracy, precision, sensitivity and f1-score.

4.1 Data Preparation and Augmentation

The use of data augmentation is a potentially useful tool in instances where there are insufficient real-world samples to facilitate the training and evaluation of the models being developed, especially in cases where a condition is rare or has only recently emerged. This process involves generating new synthetic samples based on the available data to facilitate further study [13,17,32,34].

Substitution, on the other hand, is reserved for cases where samples cannot be made public. Medical studies often face challenges due to concerns regarding data and patient privacy. As a result of these justified concerns, machine learning experts often lack access to the raw data needed for model research. Publishing patient data can also be prohibitively expensive. However, by generating fully synthetic samples that do not represent real individuals and cannot be traced back to the original source, researchers can avoid some of the stringent regulations associated with medical data, providing a starting point for developing models capable of early detection and diagnosis.

The original dataset is first divided into training and evaluation sections using a 70/30 split, which is considered standard for ML research. The training portion of the dataset is then utilized to train a GAN that can create new, fully synthetic samples based on existing inputs. A generator and discriminator network are employed to produce these new samples, with the generator creating instances to try to fool the discriminator, and the discriminator attempting to determine whether the instances are real or synthetic. Once convergence is reached between the generator and discriminator, a new set of synthetic samples is generated, roughly twice the size of the original training data. This is necessary due to an inherent limitation in GANs: if a GAN is tasked with generating a specific number of samples, the discriminator will filter out some of the generated samples, resulting in a smaller portion of output data. Random samples form the generated synthetic data are discarded until the length matches that of the original training data. The data generated from the trained GAN is then used to augment the training data for the second experiment. This is achieved by combining the existing training samples with the synthetic samples, so the augmented training dataset is twice the size of the original training data. For the third experiment, only fully synthetic samples are used to train classifiers. All evaluations are carried out using a withheld portion of testing data that was not seen by the models during training.

4.2 Real Word Data Simulations

Simulations with real world data suggest that ML algorithms are viable for tacking prediction of patient outcomes. With several of the tested algorithm exceeding 70% accuracy on the given dataset. The random forest model showcases the best performance as shown in Table 1 where and accuracy of .851852 is reported. Detail metrics shown in Table 2 show additional metrics that further highlight the classifiers good performance. As evident the random forest model once again demonstrates the best performance.

Table 1. Real world training data accuracy comparisons.

Method	Accuracy
AdaBoostClassifier	.777778
DecisionTreeClassifier	.703704
GradientBoostingClassifier	.814815
KNeighborsClassifier	.703704
MLPClassifier	.555556
RandomForestClassifier	**.851852**
SVC	.592593

Table 2. Real world training data detailed metric comparisons.

Model	Metric	Ineffective	Effective	micro Avg.	macro Avg.	weighted Avg.
AdaBoostClassifier	precision	.785714	.769231	.777778	.777473	.777778
	sensitivity	.785714	.769231	.777778	.777473	.777778
	f1-measure	.785714	.769231	.777778	.777473	.777778
DecisionTreeClassifier	precision	.750000	.666667	.703704	.708333	.709877
	sensitivity	.642857	.769231	.703704	.706044	.703704
	f1-measure	.692308	.714286	.703704	.703297	.702890
GradientBoostingClassifier	precision	.800000	.833333	.814815	.816667	.816049
	sensitivity	.857143	.769231	.814815	.813187	.814815
	f1-measure	.827586	.800000	.814815	.813793	.814304
KNeighborsClassifier	precision	.714286	.692308	.703704	.703297	.703704
	sensitivity	.714286	.692308	.703704	.703297	.703704
	f1-measure	.714286	.692308	.703704	.703297	.703704
MLPClassifier	precision	.545455	.600000	.555556	.572727	.571717
	sensitivity	.857143	.230769	.555556	.543956	.555556
	f1-measure	.666667	.333333	.555556	.500000	.506173
RandomForestClassifier	precision	.812000	.909091	.851852	.860795	.859007
	sensitivity	.928571	.769231	.851852	.848901	.851852
	f1-measure	.866667	.833333	.851852	.850000	.850617
SVC	precision	1.000000	.541667	.592593	.770833	.779321
	sensitivity	.214286	1.000000	.592593	.607143	.592593
	f1-measure	.352941	.702703	.592593	.527822	.521345
	support	14	13			

Details for the best performing model in terms of confusion matrix and RoC are provided in Fig. 2. Additionally, algorithms that demonstrate more modest performance such as the MLP and Ada boost algorithm are shown in Fig. 3. The poor performance can be associated with the relatively small size of the training data, which is not sufficient for more complex models requiring additional adjustments.

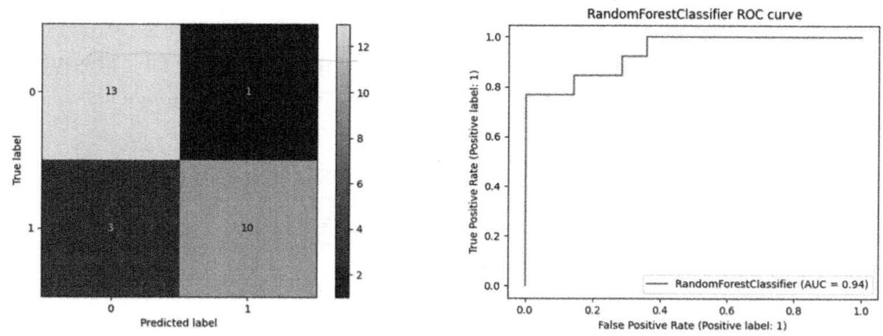

Fig. 2. Best performing model (random forest) trained on real world data confusion and RoC.

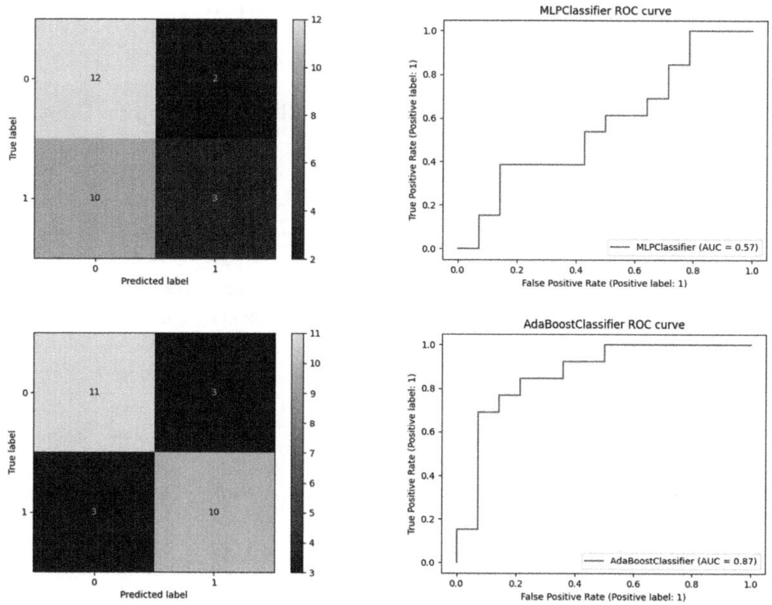

Fig. 3. MLP and AdaBoost model performance trained on real world data confusion and RoC.

4.3 Augmented Data Simulations

For data augmentation, new samples of data are generated to expand the number of training samples available. The number of samples is roughly doubled (due to the nature of GAN, the exact number of generated samples can slightly vary). Augmented data is used to train a new set of classifiers, which are then tested on real-world samples. Accuracy outcomes can be observed in Table 3. A clear improvement can be seen for the MLP and AdaBoost classifiers, which, following data augmentation, match the performance of the best algorithm. However, a reduction in the performance of the decision tree classifier and KNeighbors classifier is also observed, which may limit the applicability of the methodology in certain instances. Additional details are provided in Table 4.

Table 3. Augmented training data accuracy comparisons.

Method	Accuracy
AdaBoostClassifier	**.851852**
DecisionTreeClassifier	.629630
GradientBoostingClassifier	**.851852**
KNeighborsClassifier	.592593
MLPClassifier	**.851852**
RandomForestClassifier	**.851852**
SVC	.518519

Table 4. Augmented training data detailed metric comparisons.

Model	Metric	Ineffective	Effective	micro Avg.	macro Avg.	weighted Avg.
AdaBoostClassifier	precision	.777778	1.000000	.851852	.888889	.884774
	sensitivity	1.000000	.692308	.851852	.846154	.851852
	f1-measure	.875000	.818182	.851852	.846591	.847643
DecisionTreeClassifier	precision	.625000	.636364	.629630	.630682	.630471
	sensitivity	.714286	.538462	.629630	.626374	.629630
	f1-measure	.666667	.583333	.629630	.625000	.626543
GradientBoostingClassifier	precision	.812500	.909091	.851852	.860795	.859007
	sensitivity	.928571	.769231	.851852	.848901	.851852
	f1-measure	.866667	.833333	.851852	.850000	.850617
KNeighborsClassifier	precision	.636364	.562500	.592593	.599432	.600800
	sensitivity	.500000	.692308	.592593	.596154	.592593
	f1-measure	.560000	.620690	.592593	.590345	.589221
MLPClassifier	precision	.857143	.846154	.851852	.851648	.851852
	sensitivity	.857143	.846154	.851852	.851648	.851852
	f1-measure	.857143	.846154	.851852	.851648	.851852
RandomForestClassifier	precision	.812500	.909091	.851852	.860795	.859007
	sensitivity	.928571	.769231	.851852	.848901	.851852
	f1-measure	.866667	.833333	.851852	.850000	.850617
SVC	precision	.518519	.000000	.518519	.259259	.268861
	sensitivity	1.000000	.000000	.518519	.500000	.518519
	f1-measure	.682927	.000000	.518519	.341463	.354110
	support	14	13			

In Fig. 4, the confusion matrix and RoC plots for the MLP and AdaBoost models showcase a clear improvement compared to models trained on real data, as shown in Fig. 3. This suggests that in certain cases, data augmentation can enhance the performance of algorithms that require additional samples during the training stages.

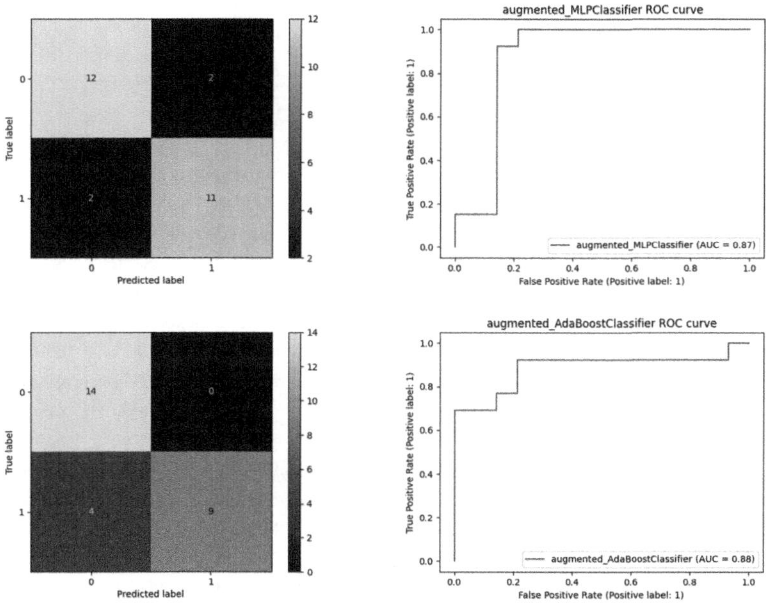

Fig. 4. MLP and AdaBoost model performance trained on augmented data confusion and RoC.

4.4 Fully Synthetic Data Simulations

In the final experiment, entirely synthetic data is employed to train models without any real samples present in the dataset. Accuracy comparisons, when tested on real samples, are presented in Table 5. Intriguingly, several algorithms once again demonstrate performance on par with the best algorithm, suggesting that synthetic data may be a viable approach for addressing this challenge. Detailed metrics are provided in Table 6. It may be observed that the results for augmented and fully synthetic data simulations are the same. The reason behind this is that the same test set was utilized for testing, to provide equal grounds for comparisons, and the same samples were problematic.

Table 5. Synthetic training data accuracy comparisons

Method	Accuracy
AdaBoostClassifier	**.851852**
DecisionTreeClassifier	.62963
GradientBoostingClassifier	**.851852**
KNeighborsClassifier	.592593
MLPClassifier	**.851852**
RandomForestClassifier	**.851852**
SVC	.518519

Table 6. Synthetic training data detailed metric comparisons.

Model	Metric	Ineffective	Effective	micro Avg.	macro Avg.	weighted Avg.
AdaBoostClassifier	precision	.777778	1.000000	.851852	.888889	.884774
	sensitivity	1.000000	.692308	.851852	.846154	.851852
	f1-measure	.875000	.818182	.851852	.846591	.847643
DecisionTreeClassifier	precision	.625000	.636364	.629630	.630682	.630471
	sensitivity	.714286	.538462	.629630	.626374	.629630
	f1-measure	.666667	.583333	.629630	.625000	.626543
GradientBoostingClassifier	precision	.812500	.909091	.851852	.860795	.859007
	sensitivity	.928571	.769231	.851852	.848901	.851852
	f1-measure	.866667	.833333	.851852	.850000	.850617
KNeighborsClassifier	precision	.636364	.562500	.592593	.599432	.600800
	sensitivity	.500000	.692308	.592593	.596154	.592593
	f1-measure	.560000	.620690	.592593	.590345	.589221
MLPClassifier	precision	.857143	.846154	.851852	.851648	.851852
	sensitivity	.857143	.846154	.851852	.851648	.851852
	f1-measure	.857143	.846154	.851852	.851648	.851852
RandomForestClassifier	precision	.812500	.909091	.851852	.860795	.859007
	sensitivity	.928571	.769231	.851852	.848901	.851852
	f1-measure	.866667	.833333	.851852	.850000	.850617
SVC	precision	.518519	.000000	.518519	.259259	.268861
	sensitivity	1.000000	.000000	.518519	.500000	.518519
	f1-measure	.682927	.000000	.518519	.341463	.354110
	support	14	13			

4.5 Best Model Interpretation

An additional contribution of this work is the feature importance analysis of the best-performing model. This interpretation helps future research identify the factors that impact model decisions. In this work, Shapley Additive Explanations (SHAP) [21] analysis is applied to the best-performing random forest model to determine the feature importance used to assess treatment outcomes. SHAP analysis can help in better comprehending of the model performance, as well as understanding of each feature's significance and impact on the final predictions made by the model. This approach is based on the Shapley values, obtained in form of a feature importance metric inspired by a game theory. The results of the analysis are shown in Fig. 5.

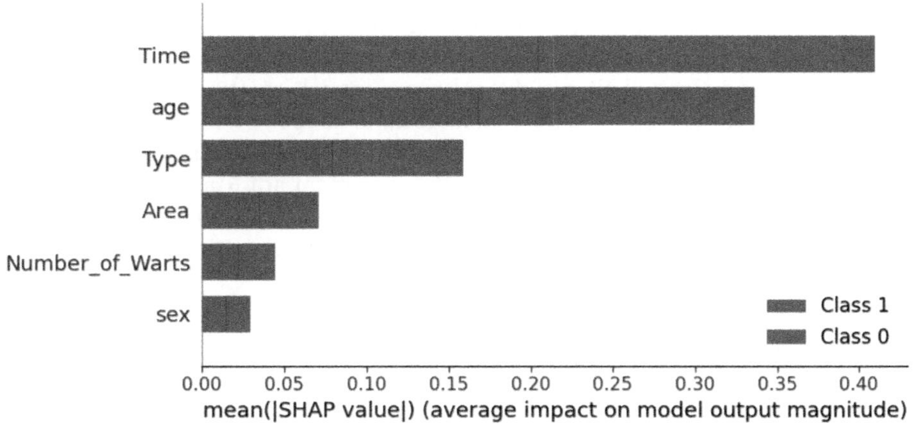

Fig. 5. Beast performing model (Random Forest) feature importance interpretations.

SHAP allocates importance to every feature, measuring its contribution on the specific forecast made by model. The analysis outcomes reveal that the number of treatments has a significant impact on patient outcomes, with patient age and the type of condition following closely in terms of importance. This suggests that the frequency of treatments is a key determinant in predicting treatment outcomes, emphasizing the relevance of considering the patient's age and the specific condition in the predictive model. Understanding these influential factors is crucial for refining the model and optimizing its predictive capabilities in the context of treatment outcomes. Feature contributions based on the best-performing model (random forest) are balanced between the two classes, meaning that all features contribute equally to a sample being assigned to a certain class. This is common with tree-based models and suggests that there should not be any bias present in the model's performance.

5 Conclusion

The use of AI emerging as a crucial tool for addressing complex and non-linear problems. Its applications, particularly in medical diagnosis and predicting treatment outcomes, showcase potential benefits for aiding professionals in decision-making. However, challenges such as algorithm selection, hyperparameter tuning, and limited medical data availability pose obstacles to seamless AI application. Simulations with real-world data affirm the viability of machine learning algorithms, with the random forest model surpassing others, achieving an accuracy of 85.19%. Notably, this work explores the use of GANs for data augmentation and synthesis, showcasing improvements in classifiers like MLP and AdaBoost, while presenting challenges for decision tree and KNeighbors models. Moreover, leveraging fully synthetic data for training yields promising results, suggesting a potential solution to address data scarcity issues. An additional contribution involves feature importance analysis, elucidating that treatment frequency significantly influences patient outcomes, underscoring the relevance of patient age and specific conditions. This insight enhances model interpretability and highlights the need for comprehensive considerations in predictive modeling.

In conclusion, this work navigates challenges, explores novel approaches, and contributes valuable insights to advance the practical application of AI, particularly in leveraging GANs for effective data augmentation and synthesis, ultimately enhancing the prediction of treatment outcomes.

References

1. Al-Antari, M.A.: Artificial intelligence for medical diagnostics-existing and future AI technology! Diagnostics **13**(4), 688 (2023)
2. Bacanin, N.: Addressing feature selection and extreme learning machine tuning by diversity-oriented social network search: an application for phishing websites detection. Complex Intell. Syst. **9**(6), 7269–7304 (2023)
3. Carreon, A., Barwey, S., Raman, V.: A generative adversarial network (GAN) approach to creating synthetic flame images from experimental data. Energy AI **13**, 100238 (2023)
4. Charbuty, B., Abdulazeez, A.: Classification based on decision tree algorithm for machine learning. J. Appl. Sci. Technol. Trends **2**(01), 20–28 (2021)
5. Creswell, A., White, T., Dumoulin, V., Arulkumaran, K., Sengupta, B., Bharath, A.A.: Generative adversarial networks: an overview. IEEE Sig. Process. Mag. **35**(1), 53–65 (2018)
6. Dobrojevic, M., Zivkovic, M., Chhabra, A., Sani, N.S., Bacanin, N., Amin, M.M.: Addressing internet of things security by enhanced sine cosine metaheuristics tuned hybrid machine learning model and results interpretation based on shap approach. PeerJ Comput. Sci. **9**, e1405 (2023)
7. Gajic, L., Cvetnic, D., Zivkovic, M., Bezdan, T., Bacanin, N., Milosevic, S.: Multilayer perceptron training using hybridized bat algorithm. In: Computational Vision and Bio-Inspired Computing: ICCVBIC 2020, pp. 689–705. Springer (2021)
8. Garcea, F., Serra, A., Lamberti, F., Morra, L.: Data augmentation for medical imaging: a systematic literature review. Comput. Biol. Med. **152**, 106391 (2023)
9. Goodfellow, I.J.: Generative adversarial nets. In: Advances in Neural Information Processing Systems, vol. 27 (2014)
10. Goodfellow, I., et al.: Generative adversarial networks. Commun. ACM **63**(11), 139–144 (2020)
11. Hastie, T., Rosset, S., Zhu, J., Zou, H.: Multi-class adaboost. Stat. Interface **2**(3), 349–360 (2009)
12. Hearst, M.A., Dumais, S.T., Osuna, E., Platt, J., Scholkopf, B.: Support vector machines. IEEE Intell. Syst. Appl. **13**(4), 18–28 (1998)
13. Jovanovic, L., Bacanin, N., Ravikumar, R., Antonijevic, M., Radic, G., Zivkovic, M.: Generative adversarial networks for synthetic training data replacement in

phishing email detection using natural language processing. In: International Conference on Smart Data Intelligence, pp. 607–618. Springer (2024)
14. Jovanovic, L., Bacanin, N., Zivkovic, M., Antonijevic, M., Petrovic, A., Zivkovic, T.: Anomaly detection in ECG using recurrent networks optimized by modified metaheuristic algorithm. In: 2023 31st Telecommunications Forum (TELFOR), pp. 1–4. IEEE (2023)
15. Jovanovic, L., Gajevic, M., Dobrojevic, M., Budimirovic, N., Bacanin, N., Zivkovic, M.: Tackling iot security challenge by metaheuristics tuned extreme learning machine. In: International Conference on Intelligent Sustainable Systems, pp. 507–522. Springer (2023)
16. Jovanovic, L., Kljajic, M., Mizdrakovic, V., Marevic, V., Zivkovic, M., Bacanin, N.: Predicting credit card churn: application of xgboost tuned by modified sine cosine algorithm. In: 2023 3rd International Conference on Smart Data Intelligence (ICSMDI), pp. 55–62. IEEE (2023)
17. Jovanovic, L., Petrovic, A., Zivkovic, T., Antonijevic, M., Bacanin, N., Zivkovic, M.: Exploring the potential of generative adversarial networks for synthetic medical data generation. In: 2023 31st Telecommunications Forum (TELFOR), pp. 1–4. IEEE (2023)
18. Jovanovic, L., Zivkovic, M., Budimirovic, N., Petrovic, A., Strumberger, I., Bacanin, N.: Metaheuristic optimized extreme gradient boosting for solar flare prediction. In: 2023 16th International Conference on Advanced Technologies, Systems and Services in Telecommunications (TELSIKS), pp. 183–186. IEEE (2023)
19. Khozeimeh, F., et al.: Intralesional immunotherapy compared to cryotherapy in the treatment of warts. Int. J. Dermatol. **56**(4), 474–478 (2017)
20. Kumar, V., Sinha, D.: Synthetic attack data generation model applying generative adversarial network for intrusion detection. Comput. Secur. **125**, 103054 (2023)
21. Lundberg, S.M., Lee, S.I.: A unified approach to interpreting model predictions. In: Guyon, I., et al. (eds.) Advances in Neural Information Processing Systems, vol. 30, pp. 4765–4774. Curran Associates, Inc., (2017)
22. Marano, G.C., Rosso, M.M., Aloisio, A., Cirrincione, G.: Generative adversarial networks review in earthquake-related engineering fields. Bull. Earthq. Eng. **22**(7), 3511–3562 (2024)
23. Margam, R.: Ethics and data privacy: the backbone of trustworthy healthcare practices. Socio-Econ. Humanistic Aspects Township Ind. **1**(2), 232–236 (2023)
24. Natekin, A., Knoll, A.: Gradient boosting machines, a tutorial. Front. Neurorobot. **7**, 21 (2013)
25. Bahri Özçelik, Y., Altan, A.: Overcoming nonlinear dynamics in diabetic retinopathy classification: a robust ai-based model with chaotic swarm intelligence optimization and recurrent long short-term memory. Fractal and Fractional **7**(8), 598 (2023)
26. Park, N., Mohammadi, M., Gorde, K., Jajodia, S., Park, H., Kim, Y.: Data synthesis based on generative adversarial networks. Proc. VLDB Endowment **11**(10), 1071–1083 (2018)
27. Peterson, L.E.: K-nearest neighbor. Scholarpedia **4**(2), 1883 (2009)
28. Podell, D., et al.: Sdxl: improving latent diffusion models for high-resolution image synthesis. arXiv preprint (2023). arXiv:2307.01952
29. Popescu, M.C., Balas, V.E., Perescu-Popescu, L., Mastorakis, N.: Multilayer perceptron and neural networks. WSEAS Trans. Circ. Syst. **8**(7), 579–588 (2009)
30. Rigatti, S.J.: Random forest. J. Insur. Med. **47**(1), 31–39 (2017)

31. Romero, H., et al.: Synthetic dataset of electroluminescence images of photovoltaic cells by deep convolutional generative adversarial networks. Sustainability **15**(9), 7175 (2023)
32. Stankovic, M., Bacanin, N., Zivkovic, M., Jovanovic, L., Sarac, M., Antonijevic, M.: Generative adversarial network for data augmentation and substitution. In: 2024 Zooming Innovation in Consumer Technologies Conference (ZINC), pp. 7–12. IEEE (2024)
33. Sun, C., van Soest, J., Dumontier, M.: Generating synthetic personal health data using conditional generative adversarial networks combining with differential privacy. J. Biomed. Inform. **143**, 104404 (2023)
34. Vega-Márquez, B., Rubio-Escudero, C., Riquelme, J.C., Nepomuceno-Chamorro, I.: Creation of synthetic data with conditional generative adversarial networks. In: 14th International Conference on Soft Computing Models in Industrial and Environmental Applications (SOCO 2019). Seville, 13–15 May 2019, Proceedings 14, pp. 231–240. Springer (2020)

Modeling and Optimization of Dynamic Systems

Optimization of Diffusion Parameters in Spherical Symmetry Models for Grain Drying

Atanas Z. Atanasov, Miglena N. Koleva(✉), and Lubin G. Vulkov

University of Ruse, 8 Studentska str., 7017 Ruse, Bulgaria
{aatanasov,mkoleva,lvalkov}@uni-ruse.bg

Abstract. In this work we consider a nonlinear parabolic equation with spherical symmetry, describing drying (shrinkage) effects of grain dehydration. The water removal is described by Fickian diffusion model and the diffusion term is dependent on the moisture content and the temperature. The formulated optimization problem for recovering two parameters in the non-linear diffusion coefficient satisfies the state constraints and minimize the cost functional, with a good accuracy, although not exactly. We develop numerical method for recovering these parameters in the non-linear diffusion term. Such a two-parameter coefficient inverse problem is ill-posed, since small errors in the input data may lead to large errors in the computed solution. The proposed approach is based on Levenberg-Marquardt method. Computational examples with synthetic and real data for chickpeas, obtained experimentally, are discussed.

Keywords: Nonlinear diffusion model · Spherical symmetry · Drying of chickpeas (Cicer arietinum) · Coefficient inverse problem · Optimization

1 Introduction

Chickpeas play an essential role in human nutrition, providing a rich source of protein, fiber, and essential nutrients. Alongside major cereal crops like wheat, corn, maize, rice, and sunflower, which provide a significant portion of world's starch needs, chickpeas contribute significantly to food security and dietary diversity worldwide, see e.g., [14,18,23].

The proper storage and processing of cereal grains are essential for maintaining their nutritional value and extending their shelf life. Effective storage techniques prevent spoilage and loss of quality due to factors like moisture, pests, and microbial contamination. Additionally, processing methods, such as protein extraction, enhance the nutritional profile of these grains by concentrating valuable nutrients. Extracted proteins from grains can be used in a variety of food products, contributing to improved dietary protein intake and supporting global food security, see e.g., [10,25,30,32].

D. Simian and L. F. Stoica (Eds.): MDIS 2024, CCIS 2486, pp. 155–169, 2025.
https://doi.org/10.1007/978-3-031-87386-7_11

Mathematical models play a crucial role in understanding and optimizing the hydration and dehydration processes of grains. These models, such as the Fickian diffusion model [13], help predict moisture transfer rates and distribution within the grains under various conditions. The movement of water within a grain, driven by a gradient in chemical potential that can be expressed through water activity, supports the application of a Fickian diffusion model to describe both water absorption and desorption. Accurate modeling is essential for designing efficient drying and hydration systems, which are vital for preserving the grains' quality and nutritional integrity. Engineering approaches and advanced simulation tools facilitate the development of these models, providing valuable insights that guide industrial practices and improve food processing technologies. Studies such as those by Page [27] and Henderson and Pabis [17] have laid the foundation for modern grain drying theories, illustrating the enduring impact of mathematical modeling in this field.

Several researchers have modeled the movement of moisture in food grains using Ficks second law of diffusion, see [10] and reference therein, where the authors provide a detailed overview of the results in this direction. Much of the research assumes that grains have a distinct geometric shape, such as a sphere or an infinite cylinder, and does not account for any changes in kernel size. Moreover, variations in volume are disregarded, and the diffusion coefficient remains constant with respect to moisture content at a specific temperature. It is well known that variable diffusivity models offer a more accurate representation of the process compared to those with a constant effective diffusivity, see e.g., [8,10,11,24,33].

It is complicated to derive an exact solution to a diffusion model when the diffusivity varies with the moisture level and thermal conditions, or when changes in size are taken into account. In [11], authors construct Crank-Nicholson finite difference scheme to address the diffusion equation with varying mass diffusivity for grains with a spherical shape. Fast solvers, based on Fourier transform, for 2D Poisson equation in polar and spherical coordinates are developed in [22].

Inverse mathematical problems are crucial for accurately modeling the hydration and dehydration processes of grains, as they allow for the determination of key parameters, such as diffusion coefficients, from experimental data. These models enable the optimization of drying and hydration processes, enhancing the efficiency and quality of grain processing. Studies such as those by [6,21,26] highlight the significant impact of inverse problem-solving in food engineering.

The model's unknown parameters in more realistic model, with a diffusion coefficient dependent on moisture content and temperature, and taking into account the growth effects caused by water absorption in white rice, are adjusted by minimizing an objective function that measures the weighted discrepancies between predictions and actual data are performed in [10].

The paper [24] proposes a moisture diffusion model for soybean grains that accounts for volume changes upon water absorption and the dependence of diffusivity on moisture content. Parametric sensitivity analysis is also performed. Parameter estimation problem in a simplified moisture uptake model is solved numerically and compared with the Fickian diffusion model in [12].

In [33], is employed a 3D scanner to capture the geometry of strawberries, pears, and potatoes, and conducted cooling-heating simulations using a one-dimensional heat conduction problem in spherical coordinates. A spherical assumption is then used to compare temperature changes at the center and average volume, using the ratios of volume to surface area to calculate their characteristic length.

In [19], various mathematical models were evaluated for the moisture absorption at various temperature settings. The authors used an artificial neural network to predict the water uptake dynamics in green chickpea. In [28], a linear model is employed to analyze water absorption in chickpeas across various temperatures and pressures. The research focuses on assessing the water diffusion rate inside the grains and calculating the energy needed for the hydration process. The processes of soaking and cooking chickpeas were investigated experimentally in [3], with regard to water uptake. A mathematical model in spherical coordinates was developed to ascertain the effective moisture diffusivity and the convective mass transfer coefficient.

Inverse problems for parabolic equations in spherical geometry are solved, for example in [4,5,15].

In this work we consider two-parameter coefficient inverse problem (2PCIP) in a nonlinear diffusion equation with spherical symmetry, proposed in [10] to model the process of water infiltration into the rice grain. We assume that the diffusion coefficient is dependent on water content and temperature and for simplicity we consider the constant grain size. Such problem is ill-posed, as even minor inaccuracies in the input data can lead to significant errors in the resulting solution, [2,9,16]. We develop numerical method for solving the 2PCIP, based on Levenberg-Marquardt algorithm. For the verification of the proposed approach, we use both synthetic and real data. The measurements are performed for Kabuli chickpeas grains in 2021/2022 in the village of Kalipetrovo, Northeastern Bulgaria.

The remainder of the paper is organized as follows. In the next section, we formulate the model problem. In Sect. 3 we propose the numerical discretizations for both direct and inverse problems. In Sect. 4 we present and discuss computational results. The paper concludes with final remarks.

2 Model Description

In the water transport characteristics study of agriculture material the moisture diffusivity is a crucial parameter. Following the Fick's second law of diffusion considering a non-constant diffusion coefficient, the moisture diffusion equation can be written in the form, see [10]:

$$\frac{\partial C}{\partial t} = \nabla.(D_0 e^{\delta C} \nabla C), \tag{1}$$

where C denotes the volumetric moisture content (kg water m^{-3} solid) in the grain (given as the ratio of water volume to wet grain volume), D_0 - adjusting parameters considering a changing diffusion coefficient (m^2s^{-1}) and δ are the parameters that must to be identified from observations of moisture diffusivity, assuming C maintains spherical symmetry throughout the volume, Eq. (1) takes the form:

$$\frac{\partial C(r,t)}{\partial t} = \frac{1}{r^2}\frac{\partial}{\partial r}\left(r^2 D_0 \exp(\delta C(r,t))\frac{\partial C(r,t)}{\partial r}\right), \tag{2}$$

This equation is defined on the domain $Q_T = \Omega \times (0,T)$, $\Omega = (0,R)$ and is associated with the subsequent initial and boundary conditions:

$$C(r,0) = C_0(r), \quad r \in \Omega, \tag{3}$$

$$C(R,t) = C_R(t), \quad t \in (0,T), \tag{4}$$

$$\frac{\partial C(0,t)}{\partial r} = 0 \ \text{ for } \ t \in (0,T), \tag{5}$$

where R denotes the radius of the grain, which is fixed.

The problem (2)–(5) is constructed under the following assumptions [10], modified for the water removal process:

- The radius of a chickpea grain ranges from about 2.5 mm to 5 mm.
- The grain material is uniform and exhibits the same properties in all directions.
- It is assumed that the grain remains at a uniform temperature throughout the drying process, which allows the omission of the heat transfer equation.
- During drying, it is assumed that the mass transfer coefficient is sufficiently high, enabling the boundary layer to gradually reach the equilibrium moisture content.
- The loss of solid mass during grain drying is considered negligible.

The problem (2)–(5) with known coefficients, right-hand side, initial and boundary conditions, is referred to as the direct (or forward) problem.

In this work, we study two-parameters coefficient inverse problem (2PCIP), namely in the problem (2)–(5) we seek δ, D_0 and solution $C(r,t)$ under additional measurements of the solution at internal points

$$C(t_l^*, r_q^*) = \psi(t^*, r^*), \quad l = 1, 2, \ldots, M_{\text{obs}}, \quad q = 1, 2, \ldots, N_{\text{obs}}. \tag{6}$$

3 Numerical Discretizations

In this section we propose numerical algorithm for solving inverse problem 2PCIP. We introduce uniform partition of the spatial and temporal interval by grid nodes $r_i = ih$, $i = 0, 1, \ldots, N$, $Nh = R$ and $t_n = n\tau$, $n = 0, 1, \ldots, M$, $M\tau = T$. The solution at grid node (r_i, t_n) is denoted by $C_i^n = C(r_i, t_n)$.

First we discuss the discretization of the direct problem.

3.1 Approximation of the Direct Problem

Following [31], we assume that

$$\lim_{r\to 0} r^2 \frac{\partial C}{\partial r} = 0, \tag{7}$$

which imply boundness of the solution $C(r,t)$. Moreover, the limit (7) yields the boundary condition (5).

Further, we apply finite volume method in space, to obtain the numerical scheme for (2)–(5). We use dual mesh

$$r_{-1/2} = r_0 < r_{1/2} < \cdots < r_{i-1/2} < r_{i+1/2} < \cdots < r_{N-1/2} < r_N = r_{N+1/2}$$

and integrate (2) over the volumes $(r_{i-1/2}, r_{i+1/2})$, $i = 1, 2, \ldots, N-1$. The resulting integral equality is

$$\int\limits_{r_{i-1/2}}^{r_{i+1/2}} r^2 \frac{\partial C(r,t)}{\partial t} = r_{i+1/2}^2 w_{i+1/2}(t) - r_{i-1/2}^2 w_{i-1/2}(t), \tag{8}$$

$$w(r,t) = D_0 \exp(\delta C(r,t)) \frac{\partial C(r,t)}{\partial r}.$$

Approximating the first integral by midpoint quadrature, replacing the fluxes w by finite differences and applying semi-implicit time stepping, we get

$$\frac{C_i^{n+1} - C_i^n}{\tau} = \frac{r_{i+1/2}^2 w_{i+1/2}^{n+1} - r_{i-1/2}^2 w_{i-1/2}^{n+1}}{h r_i^2}, \quad i = 1, \ldots, N-1, \tag{9}$$

$$w_{i+1/2}^{n+1} = D_0 \exp\left(\frac{\delta (C_{i+1}^n + C_{i-1}^n)}{2}\right) \frac{C_{i+1}^{n+1} - C_i^{n+1}}{h}.$$

At point $r = 0$, from (8), (5), we integrate in the volume $I_0 = (\varepsilon, r_{1/2} = h/2)$ to derive

$$\int\limits_{\varepsilon}^{r_{1/2}} r^2 \frac{\partial C}{\partial t} dr - \left(r_{1/2}^2 w_{i+1/2}(t) - \varepsilon^2 w_{i+1/2}(t) \right) = 0, \tag{10}$$

Using product approximation formula, we derive (see an analogous case in [31])

$$\frac{\partial C}{\partial t}(\varepsilon, t) \int\limits_{\varepsilon}^{r_{1/2}} r^2 dr - \left(r_{1/2}^2 w_{1/2}(t) - \varepsilon^2 w_{\varepsilon}(t) \right) = 0,$$

Now, letting $\varepsilon \to 0$, in view of (5), (7), for the left external boundary condition we get

$$\frac{h^3}{24} \frac{\partial C}{\partial t}(0,t) - \frac{h^2}{4} w_{1/2} = 0.$$

Approximating the derivatives by finite differences and using semi-implicit time stepping, we get

$$\frac{C_0^{n+1} - C_0^n}{\tau} = \frac{6}{h} w_{1/2}. \tag{11}$$

The approximation of Dirichlet right boundary condition (4) is straightforward

$$C_N^{n+1} = C_R(t_{n+1}). \tag{12}$$

The numerical scheme is completed by initial condition

$$C_i^0 = C_0(r_i), \quad i = 0, 1, \ldots, N. \tag{13}$$

Applying analogous considerations to those in [31, p.196], we deduce that the accuracy of the numerical scheme is $O(\tau + h^2)$.

3.2 Numerical Solution of the Inverse Problem

Let $\mathbf{p} = (\delta, D_0)$. We formulate 2PCIP in a variational form

$$\min_{\mathbf{p} \in A} F(\mathbf{p}), \quad F(\mathbf{p}) = \sum_{l=1}^{M_{\text{obs}}} \left[\sum_{q=1}^{N_{\text{obs}}} \left(C(r_q^*, t_l^*; \mathbf{p}) - \psi(r_q^*, t_l^*) \right)^2 \right]. \tag{14}$$

related to the solution of (2)–(5), where $\psi(r_q^*, t_l^*)$ are experimental data (6) and $C(r_q, t_l; \mathbf{p})$ is the solution of the problem (2)–(5).

One commonly used method for minimizing least-squares problems is the Levenberg-Marquardt method. It is a hybrid of the gradient descent and Gauss-Newton methods, acting like gradient descent when far from the optimal solution and like Gauss-Newton when close to it. Similar to the Gauss-Newton and gradient methods, it employs the same approximation for the Jacobian matrix, but distinguishes itself by utilizing a trust region strategy rather than a line search technique, see e.g. [1,2,7,9,29].

We aim to find the point $\mathbf{p}$ that is the local minimum of the functional $F(\mathbf{p})$. For this purpose, we utilize a Levenberg-Marquardt algorithm [7,20]. Equation (14) can be represented as follows

$$F(\mathbf{p}) = \mathbf{e}(\mathbf{p})^T \mathbf{e}(\mathbf{p}),$$
$$\mathbf{e} = \left[e_{11}, e_{12}, \ldots, e_{1N_{\text{obs}}}, e_{21}, \ldots, e_{2N_{\text{obs}}}, \ldots, e_{M_{\text{obs}}1}, \ldots, e_{L_{\text{obs}}N_{\text{obs}}}\right],$$
$$e_{lq} = C(r_q^*, t_l^*; \mathbf{p}) - \psi(r_q^*, t_l^*).$$

Building on the methods in [7,29], the minimization of the objective function (14) involves updating the estimated parameter vector $\mathbf{p}$ as follows:

$$\mathbf{p}^{(k+1)} = \mathbf{p}^{(k)} + \Delta\mathbf{p}^{(k+1)}.$$

Here $k = 0, 1, \ldots$ represents the iteration count and $\Delta\mathbf{p}^{(k+1)}$ is the solution of the equation

$$[J^T J + \mu \, \text{diag}(J^T J)] \Delta\mathbf{p}^{(k+1)} = J^T \mathbf{e}(\mathbf{p}^{(k)}). \tag{15}$$

In this context, μ denotes the regularization parameter that is adapted during each iteration. The term 'diag' refers to the diagonal elements, and J represents the Jacobian matrix

$$J = \begin{pmatrix} \frac{\partial e_{11}}{\partial \delta} & \frac{\partial e_{11}}{\partial D_0} \\ \frac{\partial e_{12}}{\partial \delta} & \frac{\partial e_{12}}{\partial D_0} \\ \cdots & \cdots \\ \frac{\partial e_{M_{\text{obs}} N_{\text{obs}}}}{\partial \delta} & \frac{\partial e_{M_{\text{obs}} N_{\text{obs}}}}{\partial D_0} \end{pmatrix},$$

obtained for $\mathbf{p}^{(k)}$.

The derivatives within the Jacobian matrix J are computed by

$$\frac{\partial e_{lq}}{\partial \delta} = \frac{e_{lq}(\delta + \Delta_1, D_0)}{\Delta_1}, \qquad \frac{\partial e_{lq}}{\partial D_0} = \frac{e_{lq}(\delta, D_0 + \Delta_2)}{\Delta_2}. \tag{16}$$

The termination criteria for the iterative procedure are $\left\| F\left(\mathbf{p}^{(k+1)}\right) - F\left(\mathbf{p}^{(k)}\right) \right\| \le \varepsilon$ for small $\varepsilon > 0$.

The described Levenberg-Marquardt algorithm operates as follows

Algorithm 1

1. Chose $\mathbf{p}^{(0)}$ $(k = 0)$, Δ_1, Δ_2, the precision ϵ and ε;
2. Determine J by evaluating (16) through the solution of the direct problem (9)–(13) for $\delta = p_1^{(k)}$, $D_0 = p_2^{(k)}$;
3. Determine $\Delta \mathbf{p}^{(k+1)}$ using (15);
4. Evaluate the objective function $F\left(\mathbf{p}^{(k+1)}\right)$ using (14);
5. Verify the convergence condition $\left\| F\left(\mathbf{p}^{(k+1)}\right) - F\left(\mathbf{p}^{(k)}\right) \right\| \le \varepsilon$. If it is achieved, stop the iteration process. Otherwise, set $k := k + 1$ and go to step 2.

4 Computational Results

In this section, we evaluate the effectiveness of the developed numerical approach for solving 2PCIP using Algorithm 1.

We take the data from real measurements. The experiment was conducted in 2021/2022 at the experimental field of "Agricultural Cooperative Golden Grain" in the village of Kalipetrovo, Northeastern Bulgaria. The variety grown was Kabuli, with light-colored grains and a smooth surface. During sampling, the grain moisture was checked using a METTLER TOLEDO moisture analyzer, Model: HX204, with moisture content in the individual samples ranging from 36% to 20% depending on the maturity stage. The range of measured grain moisture was sufficiently high to ensure that the chickpeas had enough moisture to investigate the drying kinetics and the change in the diffusion coefficient. To standardize the grain moisture, the samples were placed in a convective dryer at a temperature of 40°C for slower drying, which better preserves nutritional values and taste qualities. The target moisture content during drying was 15%. During drying, the change in grain moisture and water loss were monitored (causing shrinkage). Samples were periodically removed and weighed to determine the amount of evaporated water. The first change in radius is observed once the moisture content decreases to 30%. At a temperature of 40°C, this take approximately 120 to 180 min. The drying process takes approximately 300 min.

The regularization parameter μ is adapted at each iteration, following the algorithm developed in [20].

Since the time (min) is large and the diffusion coefficient D_0 is small, we introduce the rescalling $\widetilde{t} = t/100$. Thus in the normalized model the final time is reduced and $D_0 := 100D_0$. The results from simulations are presented for the original time and temperature.

Example 1 (*Discrete measurements*). We take measurements $\psi(t^*, r^*)$ in (6) at 15 random chosen grid nodes, obtained by solving numerical solution of the direct problem (9)–(13) for exact D_0 and δ.

We consider the values of variable diffusivity for dried grain (with shrinkage) $D_0 = 5.930e - 5$, $\delta = -2.451$ and variable diffusivity for undried grain (without shrinkage) $D_0 = 2.412e - 3$, $\delta = -4.32$, $C_0 = 0.28$ kg [water] kg^{-1} [dry matter], equilibrium moisture content $C_R = 0.15$ kg [water] kg^{-1} [dry matter].

To compare the values of D_0, δ and the numerical solution C, obtained by the 2PCIP with the exact values of D_0, δ (obtained experimentally) and numerical solution C of the direct problem (2)–(5), respectively, we introduce absolute and relative errors of the recovered parameters

$$E_{D_0} = |D_0 - \widetilde{D}_0|, \quad \epsilon_{D_0} = \frac{|D_0 - \widetilde{D}_0|}{D_0}, \quad E_\delta = |\delta - \widetilde{\delta}|, \quad \epsilon_\delta = \frac{|\delta - \widetilde{\delta}|}{|\delta|}.$$

Here, $\widetilde{D}_0$, $\widetilde{\delta}$ and $\widetilde{C}$ are obtained by 2PCIP, while D_0, δ and C are exact values and the numerical solution of the direct problem (2)–(5), computed for exact D_0 and δ.

In Table 1, we present the initial guess $\mathbf{p}^{(0)} = (D_0^{(0)}, \delta^{(0)})$, the absolute and relative error of the restored $\mathbf{p}$ and the number of iterations for variable diffusivity without shrinkage, $R = 5$ mm, $T = 100$ min. Results for variable diffusivity with shrinkage are presented in Table 2 for $R = 4.25$ mm, $T = 300$ min. The computations are performed for $N = 80$, $\tau = h$, $\triangle_1 = 1.e - 7$, $\triangle_2 = 1.e - 10$, $\varepsilon = 1.e - 3$.

Table 1. Absolute and relative errors of the recovered parameters for variable diffusivity without shrinkage.

$\delta^{(0)}$	$D_0^{(0)}$	$\widetilde{\delta}$	$\widetilde{D}_0$	E_δ	E_{D_0}	ϵ_δ	ϵ_{D_0}	k
-3	0.002	-4.382	2.401e$-$3	6.211e$-$2	1.132e$-$5	1.447e$-$2	4.691e$-$3	5
-2.5	0.001	-4.235	2.306e$-$3	8.501e$-$2	1.062e$-$4	1.968e$-$2	4.402e$-$2	6
-2	0.001	-4.423	2.261e$-$3	1.031e$-$1	1.510e$-$4	2.387e$-$2	6.262e$-$2	6

Table 2. Absolute and relative errors of the recovered parameters for variable diffusivity with shrinkage

$\delta^{(0)}$	$D_0^{(0)}$	$\widetilde{\delta}$	$\widetilde{D}_0$	E_δ	E_{D_0}	ϵ_δ	ϵ_{D_0}	k
-1.5	2.000e$-$5	-2.473	5.486e$-$5	2.200e$-$2	4.441e$-$6	8.975e$-$3	7.487e$-$2	5
-1	1.000e$-$4	-2.255	7.609e$-$5	1.960e$-$1	1.679e$-$5	8.000e$-$2	2.831e$-$1	7
0	1.000e$-$5	-2.546	5.704e$-$5	9.472e$-$2	2.261e$-$6	3.864e$-$2	3.813e$-$2	6

Computational results shows an optimal convergence of the recovered by 2PCIP parameters, in relation to the range size for initial approximation convergence. Additionally, it is achieved in a few iteration. Typically for Newton-like methods is the significant dependence of the appropriate choice of the initial guess.

On Fig. 1 we depict iteration process for recovering D_0 and δ for $\widetilde{\delta}^{(0)} = 0$ and $\widetilde{D}_0^{(0)} = 1.e - 3$ without shrinkage.

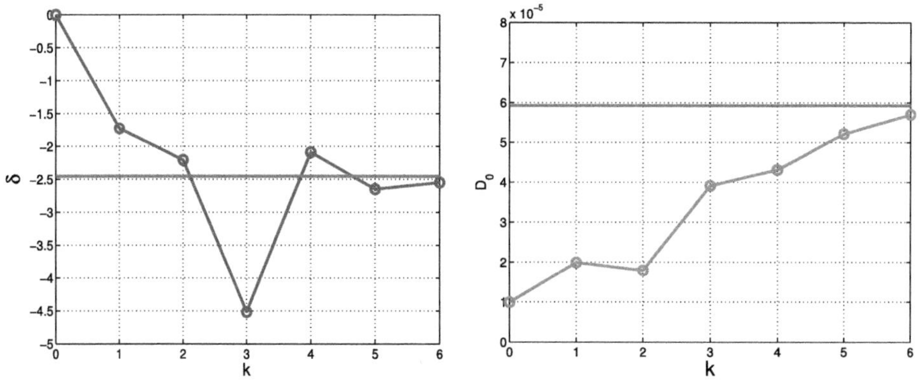

Fig. 1. Recovered (*blue line with circles*) and exact (*solid red line*) δ; recovered (*green line with squares*) and exact (*solid red line*) D_0 at each iteration. (Color figure online)

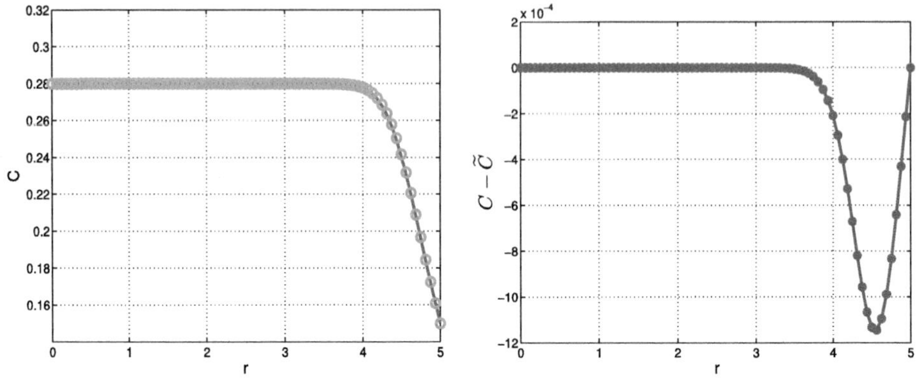

Fig. 2. Solution C, computed by direct problem (*solid red line*) and 2PCIP (*blue line with circles*) for at final time for variable diffusivity without shrinkage, $\widetilde{\delta}^{(0)} = -2.5$ and $\widetilde{D}_0^{(0)} = 0.001$ (*left*) and the corresponding error (*right*). (Color figure online)

On Fig. 2 we plot numerical solution C of the direct problem for exact δ and D_0, solution $\widetilde{C}$, computed by 2PCIP for recovered δ and D_0, at final time $T = 300$ min for variable diffusivity without shrinkage and the corresponding error.

On Fig. 3 we plot numerical solution C of the direct problem for exact δ and D_0, solution $\widetilde{C}$, computed by 2PCIP for recovered δ and D_0, at final time $T = 300$ min for variable diffusivity with shrinkage and the corresponding error.

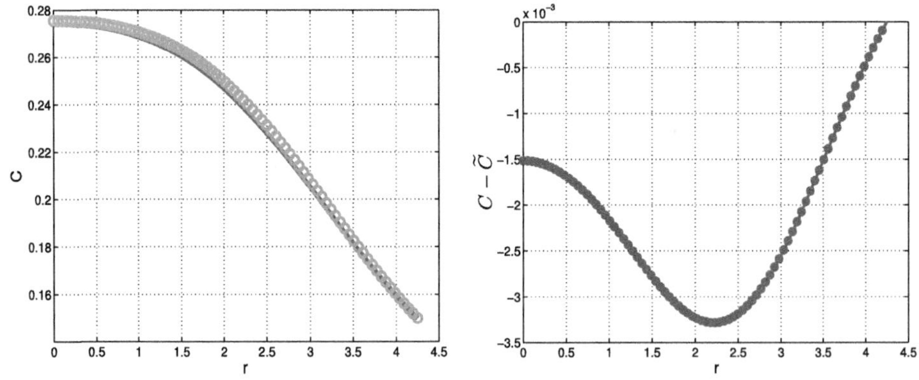

Fig. 3. Solution C, computed by direct problem (*solid red line*) and 2PCIP (*blue line with circles*) for at final time for variable diffusivity without shrinkage, $\widetilde{\delta}^{(0)} = 0$ and $\widetilde{D}_0^{(0)} = 1.000e - 3$ (*left*) and the corresponding error (*right*). (Color figure online)

Figure 2 and Fig. 3 illustrate that the parameters δ and D_0 are recovered with sufficient accuracy, such that to obtain accurate solution $\widetilde{C}$. We observe that for $T = 100$ min, moisture is removed only from the surface layer of the grain, while for $T = 300$ min, most of the grain is dried. With the considered parameters δ and D_0, more time is needed in order to dry the grain completely.

Example 2 (*Comparison*). On the base of the discretization proposed in Sect. 3.1, the 2PCIP can be solved, applying conjugate gradient method (CGM) [2,16]. In Table 3 we give computational results for variable diffusivity without shrinkage with the same mesh and model parameters as in Example 1. The results shows the advantage of the Levengerg- Marquardt algorithm in recovering the parameters δ and D_0. This influence the accuracy of the solution as well. On Fig. 4 we plot the error between the numerical solution C of the direct problem for exact δ and D_0 and the solution $\widetilde{C}$, computed by Algorithm 1 and 2PCIP with CGM, for recovered parameters δ and D_0, at final time $T = 300$ min for variable diffusivity without shrinkage.

Table 3. Absolute and relative errors of the recovered parameters for variable diffusivity without shrinkage, CGM.

$\delta^{(0)}$	$D_0^{(0)}$	$\widetilde{\delta}$	$\widetilde{D}_0$	E_δ	E_{D_0}	ϵ_δ	ϵ_{D_0}	k
−3	0.002	−4.246	2.391e−3	7.321e−2	2.101e−5	1.695e−2	8.706e−3	5
−2.5	0.001	−4.226	2.214e−3	9.414e−2	1.983e−4	1.179e−2	8.221e−2	6
−2	0.001	−4.143	2.248e−3	1.766e−1	2.537e−4	4.088e−2	1.0518e−1	7

Example 3 (*Experimental data*). The measurements are taken from experimental data. The initial and equilibrium moisture content, the final time and the radius, corresponding to the shrunk grain are the same as in Example 1. We

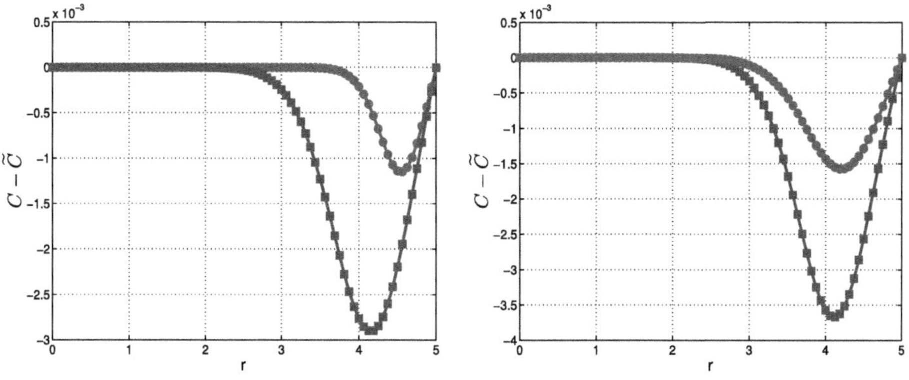

Fig. 4. Error of the solution $\widetilde{C}$, computed by Algorithm 1 (*blue line with circles*) and 2PCIP with CGM (*green line with squares*), for recovered parameters δ, D_0 and the numerical solution C of the direct problem for exact δ and D_0, at final time $T = 300$ min for variable diffusivity without shrinkage, $\widetilde{\delta}^{(0)} = -2.5$ and $\widetilde{D}_0^{(0)} = 0.001$ (*left*) and $\widetilde{\delta}^{(0)} = -2$ and $\widetilde{D}_0^{(0)} = 0.001$ (*right*). (Color figure online)

assume that the time-interpolated moisture measurements taken for different time are at the center of the chickpea grain radius. That is, we assume that the measurements are for a fixed value $r = 2.125$ mm ($N_{\text{obs}} = 1$). We use measurement points $N_{\text{obs}} = 30$.

In Table 4, we present the recovered values of δ and D_0 for different initial guess and the corresponding root mean square error (RMSE)

$$RMSE = \sqrt{\frac{\sum_{l=1}^{M_{\text{obs}}}(C_{N/2}^l - \widetilde{C}_{N/2}^l)^2}{M_{\text{obs}}}}.$$

Table 4. Recovered parameters and the corresponding RMSE for variable diffusivity with shrinkage.

$\delta^{(0)}$	$D_0^{(0)}$	$\widetilde{\delta}$	$\widetilde{D}_0$	RMSE
-4	1.000e−5	−4.523	2.836e−5	1.028e−1
-6	5.000e−3	−6.457	8.126e−3	1.001e−1
-2.5	1.000e−2	−3.031	2.442e−2	3.908e−2
-5	3.000e−2	−5.358	5.528e−2	2.311e−2

On Fig. 5, we present the solution $\widetilde{C}$ for recovered δ and D_0 in the whole computational domain and at final time $T = 300$ min.

For real data, we cannot achieve a hight precision, but the results are optimal.

The computational tests, performed in this section to test the numerical method are implemented by MATLAB® on a computer equipped with 4-cores processor and 16 GB RAM. To solve the linearized system of algebraic equations, we use direct solver, i.e. 'mldivide' toolbox.

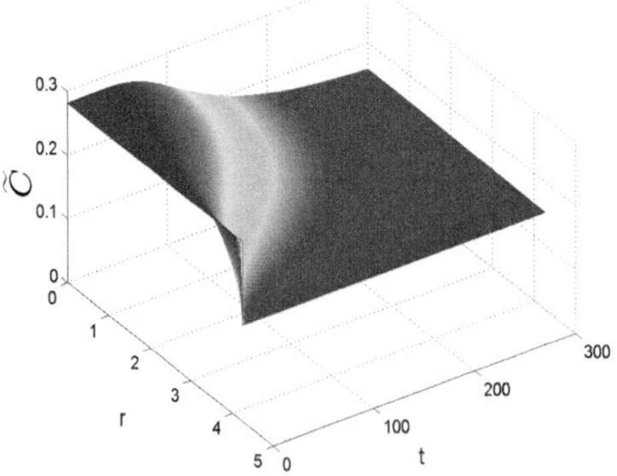

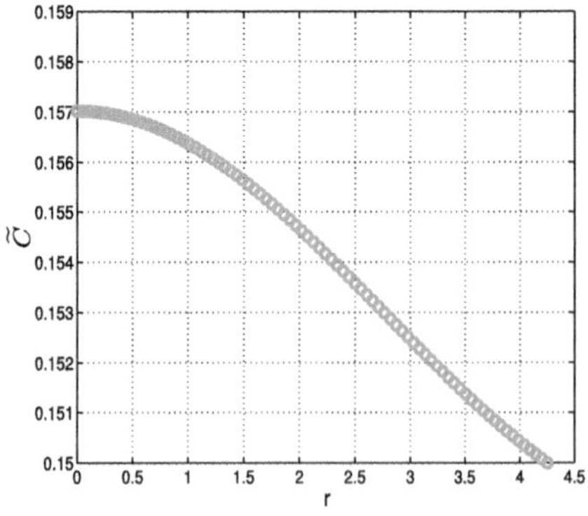

Fig. 5. Solution $\widetilde{C}$, computed for recovered parameters δ and D_0 in the whole computational domain (*left*) and at final time $T = 300$ min (*right*).

5 Conclusions

The key findings of this study can be outlined as follows.

- We propose a finite difference method for solving of a parabolic equation with an exponentially nonlinear diffusion coefficient and spherical symmetry. It describes drying (shrinkage) effects of grain dehydration.
- Following agriculture practice, we formulate an optimization inverse problem to recover two parameters in the nonlinear diffusion coefficient that are difficult to measure. We develop a numerical method for minimizing the corresponding cost functional.
- Test numerical examples with synthetic and experimentally obtained real data for chickpeas are analyzed. Results from numerical experiments showed that the convergence is achieved in just a few iterations. For syntectic data, we obtain a good precision of the numerical solution computed with the recovered parameters. We observe good convergence of the inverse problem solution concerning the size of the convergence range for the initial guess Simulations with real data provide sufficiently accurate results.

In our future work, we plan to consider a more realistic case of varying the grain radius while accounting for the drying shrinkage effect, i.e., inverse problem with moving boundary.

Acknowledgments. The first author is supported by the project No. 24-FAI-01, financed by the "Scientific Research" fund of the University of Ruse. The second and the third authors are supported by the Bulgarian National Science Fund under the Project KP-06-N 62/3 "Numerical methods for inverse problems in evolutionary differential equations with applications to mathematical finance, heat-mass transfer, honeybee population and environmental pollution", 2022.

References

1. Aster, R.C., Borchers, B., Thurber, C.H.: Parameter Estimation and Inverse Problems, 2nd edn. Academic Press, Elsevier, USA (2013)
2. Beck, J.V., Arnold, K.J.: Parameter Estimation in Engineering and Science. Wiley, New York (1977)
3. Bidkhori, P., Karizaki, V.M.: Diffusion and kinetic modeling of water absorption process during soaking and cooking of chickpea. Legume Sci. **4**(1), 1–11 (2022)
4. Cheng, W., Fu, C.L., Quan, A.: A modified Tikhonov regularization method for a spherical symmetric three-dimensional inverse heat conduction problem. Math. Comput. Simul. **75**, 97–112 (2007)
5. Cheng, W., Liu, Y.L., Yang, F.: A modified regularization method for a spherically symmetric inverse heat conduction problem. Symmetry **14**, 2102 (2022)
6. Crank, J.: The Mathematics of Diffusion, 2nd edn. Cladendon Press, Oxford (1975)
7. Cui, M., Yang, K., Xu, X., Wang, S., Gao, X.: A modified Levenberg-Marquardt algorithm for simultaneous estimation of multi-parameters of boundary heat flux by solving transient nonlinear inverse heat conduction problems. Int. J. Heat Mass Transf. **97**, 908–916 (2016)

8. Da Silva, W.P, e Silva, M.D.P.S., da Silva, L.D, Lins, M.A.A.: Comparison between models with constant and variable diffusivity to describe water absorption by composite materials. Materialwissenschaft und Werkstofftechnik **43**(10), 825–831 (2012)
9. Dostert, P., Efendiev, Y., Mohanty, B.: Efficient uncertainty quantification techniques in inverse problem for Richards equation using coarse-scale simulation models. Adv. Water Resour. **32**(3), 329–339 (2009)
10. Dutta, A., Subramanian, A.S., Chakraberty, R.: Numerical modeling of water uptake in white rice (Oryza sativa) using variable diffusivity approach. Biosyst. Eng. **191**, 116–128 (2020)
11. Dutta, S.K., Nema, V.K., Bhardwaj, R.K.: Drying behavior of spherical grains. Int. J. Heat Mass Transf. **31**(4), 855–861 (1988)
12. Dutta, A., Chanda, A., Chakraborty, R.: A linear driving force (LDF) approximation of moisture diffusion kinetics in white rice. Int. J. Food Eng. **4**(8), Article 2 (2008)
13. Fick, A.: Ueber diffusion. Ann. Phys. **170**(1), 59–86 (1855)
14. Foyer, C., Lam, H.M., Nguyen, H., et al.: Neglecting legumes has compromised human health and sustainable food production. Nat. Plants **2**, 16112 (2016)
15. Geng, X., Cheng, H., Liu, M.: Inverse source problem of heat conduction with time-dependent diffusivity on a spherical symmetric domain. Inverse Probl. Sci. Eng. **29**(11), 1–16 (2021)
16. Hasanov, A.H., Romanov, V.G.: Introduction to Inverse Problems for Differential Equations, 1st edn., p. 261. Springer, Cham (2017)
17. Henderson, S.M., Pabis, S.: Grain Drying Theory. I.: temperature effect on drying coefficient. J. Agric. Eng. Res. **6**(3), 169–174 (1961)
18. Jukanti, A.K., Gaur, P.M., Gowda, C., Chibbar, R.N.: Nutritional quality and health benefits of chickpea (Cicer arietinum L.): a review. Br. J. Nutr. **108**(S1), S11–S26 (2012)
19. Kumar, Y., Singh, L., Sharanagat, V.S., Tarafdar, A.: Artificial neural network (ANNs) and mathematical modelling of hydration of green chickpea. Inf. Process. Agric. **8**(1), 75–86 (2021)
20. Kwak, Y., Hwang, J., Yoo, C.: A new damping strategy of Levenberg-Marquardt algorithm for multilayer perceptrons. Neural Netw. World **21**(4), 327–340 (2011)
21. Lahsasni, S., Kouhila, M., Mahrouz, M., Jaouhari, J.T.: Drying kinetics of prickly pear fruit (Opuntia ficus indica). J. Food Eng. **61**(2), 173–179 (2004)
22. Lai, M.C., Wang, W.C.: Fast direct solver for Poisson equation on 2D polar and spherical geometries. Numer. Meth. Partial Differ. Equ. **8**(1), 56–68 (2002)
23. Mudryj, A.N., Yu, N., Aukema, H.M.: Nutritional and health benefits of pulses. Appl. Physiol. Nutr. Metab. **39**(11), 1197–1204 (2014)
24. Nicolin, D.J., Jorge, R.M., Jorge, L.: Evaluation of distributed parameter mathematical models applied to grain hydration with volume change. Heat Mass Transf. **51**, 107–116 (2015)
25. Oghbaei, M., Prakash, J., Yildiz, F.: Effect of primary processing of cereals and legumes on its nutritional quality: a comprehensive review. Cogent Food Agric. **2**(1), 1136015 (2016)
26. Onwude, D.I., Hashim, N., Janius, R.B., Nawi, N.M., Abdan, K.: Modeling the thin-layer drying of fruits and vegetables: a review. Compr. Rev. Food Sci. Food Saf. **15**(3), 599–618 (2016)
27. Page, G.E.: Factors Influencing the Maximum Rates of Air Drying Shelled Corn in Thin Layers. M.S. thesis, Purdue University (1949)

28. Pramiu, P.V., Rizzi, R.L., do Prado, N.V., Coelho, S.R.M., Bassinello, P.Z.: Numerical modeling of chickpea (Cicer arietinum) hydration: the effects of temperature and low pressure. J. Food Eng. **165**, 112–123 (2015)
29. Rodriguez-Quinonez, J.C., Sergiyenko, O., Gonzalez-Navarro, F.F., Basaca- Preciado, L., Tyrsa, V.: Surface recognition improvement in 3D medical laser scanner using Levenberg-Marquardt method. Signal Process. **93**, 378–386 (2013)
30. Rachwa-Rosiak, D., Nebesny, E., Budryn, G.: Chickpeas-composition, nutritional value, health benefits, application to bread and snacks: a review. Crit. Rev. Food Sci. Nutr. **55**(8), 11371145 (2015)
31. Samarskii, A.: The Theory of Difference Schemes, 1st edn. Marcel Dekker, New York (2001)
32. Tang, J., Yao, D., Xia, S., Cheong, L., Tu, M.: Recent progress in plant-based proteins: from extraction and modification methods to applications in the food industry. Food Chem. X **23**, 101540 (2024)
33. Uyar, R., Erdoglu, F.: Numerical evaluation of spherical geometry approximation for heating and cooling of irregular shaped food products. J. Food Sci. **77**, E166–E175 (2012)

Parameter Optimization in a Model of Influence of Brood Deaths on Honeybee Population Dynamics

Atanas Z. Atanasov[1], Slavi G. Georgiev[2,3](✉), and Lubin G. Vulkov[3]

[1] Department of Agricultural Machinery, Agrarian-Industrial Faculty, University of Ruse, Ruse, Bulgaria
aatanasov@uni-ruse.bg

[2] Department of Information Modeling, Institute of Mathematics and Informatics, Bulgarian Academy of Sciences, Sofia, Bulgaria
sggeorgiev@math.bas.bg

[3] Department of Applied Mathematics and Statistics, Faculty of Natural Sciences and Education, University of Ruse, Ruse, Bulgaria
{sggeorgiev,lvalkov}@uni-ruse.bg

Abstract. Numerous studies have investigated the various stressors that can cause the collapse of honeybee colonies. Given the challenges in directly observing honeybee population dynamics, it is essential to construct mathematical models to gain insight into the factors that significantly affect bee populations. Several models have been proposed to capture the dynamics of honeybee populations and the effects of human-induced stressors, such as pathogens, parasites, and nutritional deficiencies, on the colony's long-term survival. However, these models generally overlook the death rate of brood, concentrating instead on the mortality rates of foraging bees. We consider a model of a nonlinear system of ordinary differential equations with main unknowns the brood B and the hive bees H. We formulate an optimization problem for recovering coefficients in the system that satisfies the state constraints and minimize the cost functional. The nature of the algorithm is to numerically reconstruct the unknown parameters, i. e. they are fitted with a good accuracy, although not exactly. We develop a computational approach to solve the problem. Numerical experiments with synthetic and real data are analyzed.

Keywords: Honeybee Population Dynamics · Colony Collapse Disorder · Brood Death · Parameter Identification · Adjoint Equations

1 Introduction

A honey bee colony is composed of several castes, including brood (eggs and larvae), sealed brood (pupae), queens, male drones, and female workers. The primary function of drones is to mate with a queen during seasonal nuptial flights, after which they die. The queen's main role is egg production. Upon emerging from her cell, a virgin queen searches for other virgin queens to eliminate them and subsequently seeks a mate to

D. Simian and L. F. Stoica (Eds.): MDIS 2024, CCIS 2486, pp. 170–182, 2025.
https://doi.org/10.1007/978-3-031-87386-7_12

commence egg-laying. The development cycle from egg to adult bee spans approximately 21 d, culminating in the emergence of adult workers. As these workers age, their responsibilities transition from tasks within the hive to those outside it. Adult workers engage in various activities, including brood care, hive maintenance, and foraging [17]. Honey bees generate sustenance for themselves, the queen, and their brood. In addition to honey production, they play a crucial role in the pollination of numerous plant species.

The population dynamics of honey bees exhibit a seasonal pattern driven by resource availability. During the spring months, extended daylight hours and increased pollen availability lead to a birth rate that surpasses the death rate, causing the bee population to expand. The population typically peaks around late June and is still elevated until mid-summer, after which it begins to decrease. This decline continues through fall and winter, reaching its lowest point in February. Concerning brood dynamics, the queen's egg-laying rate is minimal during late winter but accelerates rapidly in spring to capitalize on the blooming period. Although this rate remains high during summer, it is lower than spring levels. A secondary peak in egg-laying may occur in the fall due to the blooming of certain plant species like ironweed. After this brief resurgence, the egg-laying rate declines, approaching zero by December [18].

Honey bees not only produce honey for their consumption but also actively participate in the pollination of various plants, which is vital for ecological balance and agricultural productivity. The intricate dynamics of honey bee populations underscore the importance of understanding factors influencing their survival and growth, particularly in light of increasing anthropogenic stressors. This study aims to enhance existing mathematical models by incorporating brood death rates, offering a more comprehensive understanding of honey bee population dynamics and aiding in the development of effective conservation strategies.

Numerous studies have been conducted to identify the types of stressors that lead to the collapse of honey bee colonies, a phenomenon that has been observed almost annually for over a century [6,7,23]. This event, often referred to as Colony Collapse Disorder (CCD), has been documented across various regions of the world, affecting both wild and domesticated honeybee populations [22]. The potential causes for colony failure are varied and multifaceted, including biological factors such as parasites and pathogens, as well as anthropogenic influences like agricultural pesticides, beekeeper-applied chemicals, genetically modified crops, and insecticides. Changes in cultural practices related to agriculture and land use have also contributed to altering the natural habitats of honeybees, exacerbating the stressors that these pollinators face [19,20].

Given the complexity and challenges associated with directly observing honeybee population dynamics, developing mathematical models becomes essential for understanding the factors that significantly influence bee populations. Several mathematical models have been formulated to describe honey bee population dynamics and the impact of anthropogenic stressors–such as pathogens, parasites, and nutritional stressors–on the long-term viability of colonies [3,4,9–11,14].

Among the earlier research, significant contributions include the work of Khoury et al. (2011), who introduced a model to explore varying forager mortality rates and their effects on colony growth and development [12,13]. Russell et al. (2013) examined the sensitivity of a colony by considering seasonal fluctuations, forager mortality, food availability, and factors that influence the transition age from worker to forager, utilizing discrete data [21]. Perry et al. (2015) carried out a case study to gather data on the consequences of food shortages and the effects of accelerated maturation on the success of honeybee colonies [20].

Despite the valuable insights these models offer, they all overlook the death rate of brood and instead concentrate on the mortality of foraging bees. As a result, none of these models directly explore how brood mortality might lead to colony collapse. The study by [5] introduces a mathematical model that assesses the impact of brood death rates–caused by insecticide exposure–on the dynamics of honeybee populations, while also accounting for the seasonal variations in the egg-laying rate.

This gap in the existing literature highlights the need for a comprehensive approach that includes brood death rates in the models. By doing so, we can achieve a more accurate understanding of the factors contributing to colony collapse and develop more effective strategies for preserving honey bee populations.

The theoretical part of the paper studies the stability of the brood honeybee model described in [5] and investigates how the brood delay τ affects the population dynamics in the hive.

We first study local stability of the equilibrium points of the ODEs. Next, using the delay τ as bifurcation parameter, we that the positive equilibrium loses it's stability and the equation exhibits Hopf bifurcation.

Our goal is to use the theoretical results for this model to obtain realistic predictions for honeybee population levels and to examine the dynamics of honeybee colonies in villages Brestovitza, Yuper, etc. [1].

The remaining of the paper is structured as follows. In the next section, the models are thoroughly described from a biological point of view. Then, the relevant inverse problems are defined and the adjoint equations optimization approach is applied in Sect. 3. Section 4 is dedicated to extensive numerical experiments, while the last section concludes the paper.

2 Model Formulation

Among the previous studies, such as the review by [2], one of the most significant contributions is from [13], which focuses primarily on modeling the impact of food on hive dynamics.

In this paper, we focus on a simpler model to analyze in detail the effect of food delays on hive population dynamics.

2.1 Biological Assumptions

Let $B(t)$ represent the number of brood (eggs, larvae, pupae) in a given hive at time t, and let $H(t)$ denote the number of adult bees in the hive at time t. The time t is measured in units of 12 d, as data is collected at this interval. The ecological assumptions for the population dynamics of Apis mellifera are as follows:

- The brood are born into the brood compartment at an egg-laying rate r by the queen. It is assumed that the queen is fertile and always well-fed.
- The brood are cared for by the adult bees according to a Holling Type 3 functional response. This response is chosen because a minimum number of adults, K, is needed to care for the brood. If there are insufficient adult bees, the colony will head towards extinction, but if there are enough, the colony will grow.
- Brood mortality is assumed to follow an exponential decay, proportional to the number of brood present at time t. Similarly, adult bee mortality is also exponential, proportional to the number of adults present at time t.
- Brood transition into adults at a constant rate ϕ, referred to as the **eclosion rate**.
- The model assumes the absence of drones, as their primary function is to reproduce with other queens.

The Holling Type 3 functional response, proposed by Holling in 1975, is used to model the "birth" of brood. This response accounts for the **capture** rate, which increases in direct proportion to the adult bee population. At low adult population levels, the rate of brood **capture** accelerates. Let t_s denote the available **capture** time, which is the time during which adults can search for and care for the brood.

These assumptions lead to the following system of nonlinear ordinary differential equations (ODEs) modeling the population dynamics of Apis mellifera.

2.2 The ODEs Model

In this model, r represents the egg-laying rate, and K denotes the critical population of adult bees over time. K is considered the critical number of bees required to care for the brood.

The adult bees available must locate and tend to more brood to ensure the hive's success. Based on aforementioned assumptions, the authors of [5] proposed the following model:

$$\begin{aligned} \frac{dB}{dt} &= \frac{r(t)H^2}{K^2+H^2} - \mu_b B - \phi B \equiv f(B,H), \\ \frac{dH}{dt} &= \phi B - \mu_h H \equiv g(B,H) \end{aligned} \tag{1}$$

A mathematical analysis of model (1) is presented in [5, Section 3].

It is demonstrated that the solutions of the system (1) are bounded and positive, and (1) has no limit cycles in $\mathbb{R}^2$.

Therefore, the system (1) is biologically well-defined, ensuring that the population remains bounded over time. The upper limit of each compartment signifies the maximum possible populations of brood and adult bees.

For the brood population, this upper limit is directly related to the egg-laying rate r. A lower value of r results in a smaller upper bound for the brood population. Likewise, if the combined rate of brood mortality and maturation, $\mu_b + \phi$, is excessively high, the upper limit for the brood population will also be reduced.

Therefore, for practical applications that assist beekeepers, it is necessary to estimate the parameters of the system (1) based on concrete measurements. Similar studies have been conducted in papers [1,5].

3 Parameter Optimization

In this section, we formulate inverse problems that aim to provide realistic predictions for identifying honeybee population dynamics.

To obtain estimates for model parameters that are challenging to measure directly, we optimize the following problems.

3.1 Inverse Problems

The coefficients in the system (1), based on previous studies and beekeepers' experience [5, 12, 13], are assumed to be nonnegative.

In practical beekeeping, some of these coefficients are not directly observable and must be approximated using available measured data on the number of brood (eggs, larvae, pupae) $B(t)$ and the number of adult bees $H(t)$ in a given hive at time t.

Honeybee practice necessitates the formulation of two inverse problems, IP1 and IP2.

IP1. Following the analysis of synthetic and real data and biological considerations [24], we investigate the constant coefficients $p \equiv (p^1, p^2, p^3),\ p^1 = \mu_b,\ p^2 = \phi,\ p^3 = \mu_h$ to be difficult to measure directly.

We explore the inverse problem of reconstructing the parameters

$$p \in P_{adm} = \{0 < p^i < p_{max},\ i = 1,2,3\} \tag{2}$$

through the observed behavior

$$U(t^i) := \{\mathscr{B}(t^i) = \mathscr{B}^i,\ i = 1,\dots,I_B\ \ \mathscr{H}(t^i) = \mathscr{H}^i,\ i = 1,\dots,I_H \\ t_0 = t^1 < \cdots < t^{I_{obs}} = T,\ I_{obs} = \max(I_B, I_H)\} \tag{3}$$

of the dynamic system (1).

IP2. Now we consider all the constant coefficients

$$p = (p^1, p^2, p^3, p^4, p^5),\ p^1 = \mu_b,\ p^2 = \phi,\ p^3 = \mu_h,\ p^4 = K^2,\ p^5 = r \tag{4}$$

through the same observed behavior (3).

3.2 Optimization Approach

We consider in detail the IP1.

We solve the point observation problem (1)–(3) by minimizing an appropriate least-square functional [15, 16].

This optimization approach allows us to estimate the difficult-to-measure parameters of the system (1) based on observed data, thereby enhancing the accuracy and applicability of our model in predicting honeybee population dynamics.

We are going to minimize the functional

$$J(p) = J(\mu_b, \phi, \mu_h) = J_B(p) + J_H(p), \tag{5}$$

where

$$J_B(p) = \sum_{i=1}^{I_B} (B(t^i; p) - \mathscr{B}^i)^2, \; J_H(p) = \sum_{i=1}^{I_H} (H(t^i; p) - \mathscr{H}^i)^2. \tag{6}$$

Theorem 1. *The gradient $J'_p \equiv \{J'_{\mu_b}, J'_\phi, J'_{\mu_h}\}$ of the functional $J(p)$ is given by:*

$$J'_{\mu_b}(\mathbf{p}) = \int_0^T B\varphi_B dt, \tag{7}$$

$$J'_\phi(\mathbf{p}) = \int_0^T B(\varphi_B - \varphi_H) dt, \tag{8}$$

$$J'_{\mu_h}(\mathbf{p}) = \int_0^T H\varphi_H dt, \tag{9}$$

where the pair of the functions $\{\varphi_B = \varphi_B\ (t),\ \ \varphi_H = \varphi_H\ (t)\}$ are the unique solution to the adjoint final value problem

$$\frac{d\varphi_B}{dt} = (\mu_b + \phi)\varphi_B - \phi\varphi_H + 2\sum_{i=1}^{I_B} (B(t; \mathbf{p}) - \mathscr{B}(t))\delta(t - t^i), \tag{10}$$

$$\frac{d\varphi_H}{dt} = -2\frac{rK^2H}{(K^2 + H^2)^2}\varphi_B + \mu_h\varphi_H + 2\sum_{i=1}^{I_h} (H(t; \mathbf{p}) - \mathscr{H}(t))\delta(t - t^i), \tag{11}$$

and $\mathscr{B}(t), \mathscr{H}(t)$ are interpolants of the discrete functions taking value $\mathscr{B}(t^i)$, $i = 1, \ldots, I_B$ and $\mathscr{H}(t^i)$, $i = 1, \ldots, I_H$, respectively.

Proof (Outline). We denote $\delta\mathbf{p} = (\delta\mu_b, \delta\phi, \delta\mu_h)$, $\delta\mu_b = \varepsilon h_1$, $\delta\phi = \varepsilon h_2$, $\delta\mu_h = \varepsilon h_3$ and $\delta B(t; \mathbf{p}) = B(t; \mathbf{p} + \delta\mathbf{p}) - B(t; \mathbf{p})$, $\delta H(t; \mathbf{p}) = H(t; \mathbf{p} + \delta\mathbf{p}) - H(t; \mathbf{p})$.

Now, we rewrite the system (1) at $\mathbf{p} := \mathbf{p} + \delta\mathbf{p}$ for the pair $\{B(t; \mathbf{p} + \delta\mathbf{p}), H(t; \mathbf{p} + \delta\mathbf{p})\}$ with initial value B_0, H_0, respectively.

Next, we perform the differences between the corresponding equations to obtain a system of ordinary differential equations for the pair $\{\delta B(t; \mathbf{p}), \delta H(t; \mathbf{p})\}$ with zero initial conditions, which reads:

$$\frac{d}{dt}\delta B = -(\mu_b + \phi)\delta B + 2\frac{rK^2H}{(K^2 + H^2)^2}\delta H - \delta\mu_b B - \delta\phi B + \mathscr{O}(\varepsilon),$$

$$\frac{d}{dt}\delta H = \phi\delta B - \mu_h\delta H + \delta\phi B - \delta\mu_h H + \mathscr{O}(\varepsilon).$$

On the second stage, we calculate the increment of the functional $J(\mathbf{p})$ to obtain:

$$\begin{aligned} J(\mathbf{p} + \delta\mathbf{p}) - J(\mathbf{p}) &= 2\sum_{i=1}^{I_B} \int_0^T \delta B(t; \mathbf{p})(B(t; \mathbf{p}) - \mathscr{B}(t))\delta(t - t^i) dt \\ &\quad + 2\sum_{i=1}^{I_H} \int_0^T \delta H(t; \mathbf{p})(H(t; \mathbf{p}) - \mathscr{H}(t))\delta(t - t^i) dt + \mathscr{O}(\varepsilon). \end{aligned} \tag{12}$$

Following the technology of [15,16], we multiply the equations for $\{\delta\ \mathrm{B}\ (t;\mathbf{p}\ ),$ $\delta\ \mathrm{H}\ (t;\mathbf{p}\ )\}$ by smooth functions $\varphi_B(t)$ such that $\varphi_B(T)=0,\ \varphi_H(t)$ with $\varphi_H(T)=0$. We integrate both sides of the results from 0 to T and add them together. Also, using the expressions for $\frac{d\varphi_B}{\mathrm{dt}}, \frac{d\varphi_H}{\mathrm{dt}}$ from (10), (11) and employing (12), after some algebra we find

$$
\begin{aligned}
J(\mathbf{p}+\delta\mathbf{p})-J(\mathbf{p}) &= J(\mu_b+\varepsilon h_1,\ \phi+\varepsilon h_2,\ \mu_h+\varepsilon h_3)-J(\mu_b,\phi,\mu_h)\\
&= -\int_0^T \varphi_B B\delta\mu_B dt-\int_0^T \varphi_B B\delta\phi dt+\int_0^T \varphi_H B\delta\phi dt-\int_0^T \varphi_H H\delta\mu_h dt.
\end{aligned}
$$

Now, taking $h_2=h_3=0$, dividing the both sides of the last expression by εh_1 and passing to the limit $\varepsilon\to 0$, we obtain the formula for J'_{μ_b} in (7).

In the same way one can check the validity of the other formulas in (7)–(9). □

3.3 Algorithm

We now give the computational algorithm to solve IP1 (1)–(3). We need to find those values $\check{\boldsymbol{p}}$, called minimizer, which minimizes the functional (5), see Algorithm 1 and Fig 1.

Algorithm 1. Adjoint Equation Optimization Method

Initialize $\boldsymbol{p}_0\in\mathbb{S}_{\mathrm{adm}}$ and $l=-1$.
repeat
  $l:=l+1$
  Solve the direct problem (1),(2) at the current value $\boldsymbol{p}_l$. Then, define $U(t^i)$ as in (3)
  Solve the adjoint problem (10),(11)
  Compute $J'(\boldsymbol{p}_l)$ by formulas (7)-(9)
  Using a conjugate gradient method, define the optimization parameter $\boldsymbol{r}_l>0$ and compute the new parameter value $\boldsymbol{p}_{l+1}$ by formula

$$
\boldsymbol{p}_{l+1}=\boldsymbol{p}_l-\boldsymbol{r}_l J'(\boldsymbol{p}_l),\quad \boldsymbol{r}_l>0,\ \boldsymbol{r}_l\in\mathbb{R}_3^+
$$

until $\|\triangle\boldsymbol{p}_l\|:=\|\boldsymbol{p}_{l+1}-\boldsymbol{p}_l\|<\varepsilon_{\boldsymbol{p}}$
Set $\check{\boldsymbol{p}}:=\boldsymbol{p}_{l+1}$

The threshold $\varepsilon_{\boldsymbol{p}}$ is chosen small, for example $\varepsilon_{\boldsymbol{p}}=1e-16$ and $\boldsymbol{r}_l$ are determined empirically.

4 Computational Simulations

In this section, we perform numerical simulations to verify the theoretical results.

4.1 Forward Problem

In this subsection, we test model (1) with data from [8] and set as follows. For the 1975 data, we use $r = 25061.572$, $\mu_B = 0.575$, $\mu_H = 0.302$, $\mu_B = 0.571$, $K = 3805.467$ and the initial conditions are $B_0 = 6125, H_0 = 5362$ [5].

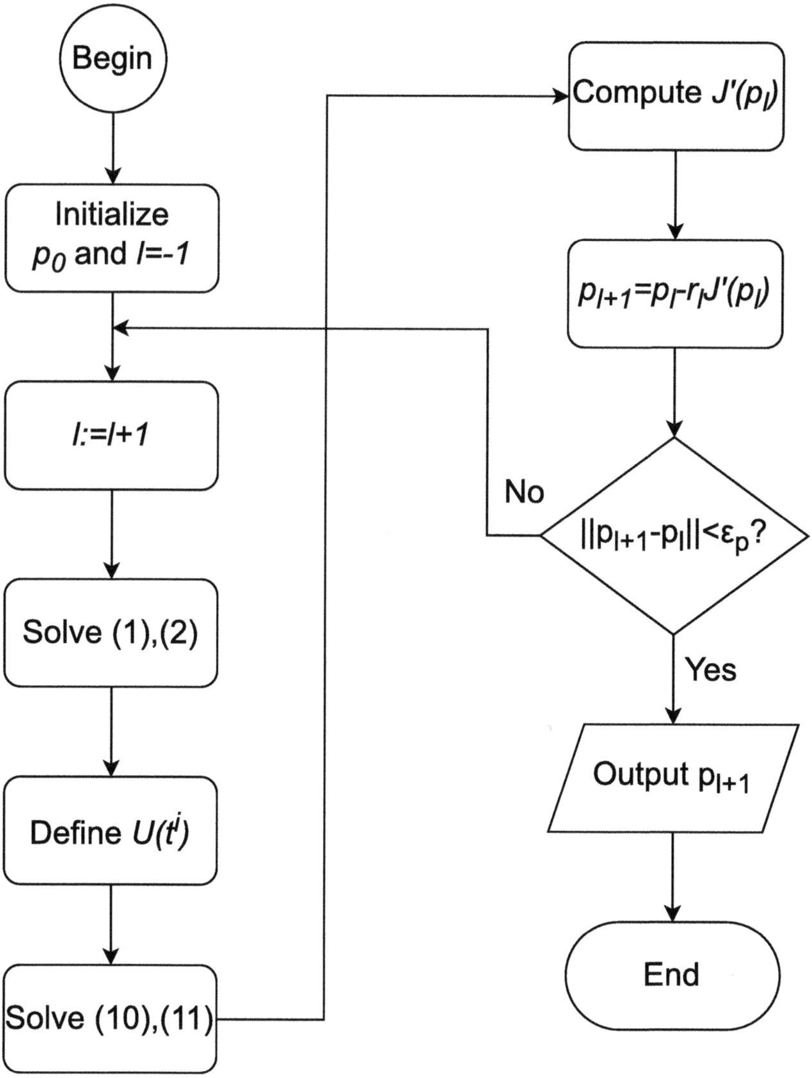

Fig. 1. The flowchart of Algorithm 1.

For the 1976 data, we employ $r = 27666.177$, $\mu_B = 0.583$, $\mu_H = 0.246$, $\phi = 0.571$, $K = 4732.389$ and the initial conditions are $B_0 = 5982, H_0 = 5362$ [5]. The results are plotted on Fig. 2.

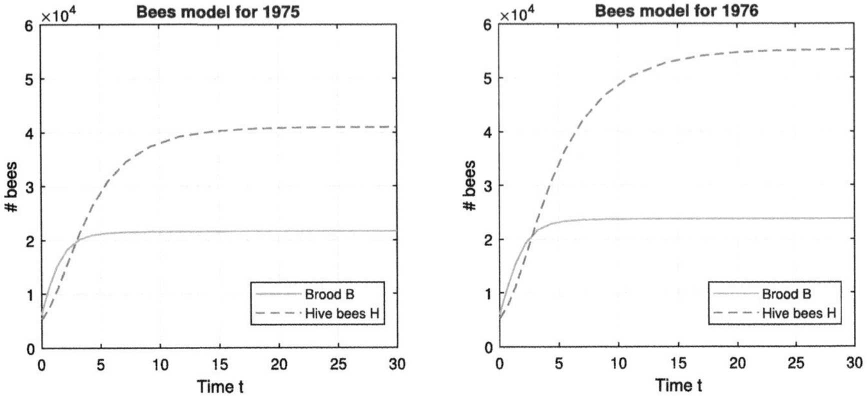

Fig. 2. The solution to (1) for 1975 data (left) and 1976 data (right).

4.2 Inverse Problem

Here we will present results concerning the inverse problem IP1. Let us again use the data for 1975. We aim to recover the parameter set $\boldsymbol{p} = (\mu_B, \phi, \mu_H)^\top = (0.575, 0.571, 0.302)^\top$ solving (1)–(3). We look for the parameter values in the unit cube $\boldsymbol{p} \in \$\text{adm} \equiv (0,1)^3$ with tolerance $\varepsilon\delta = 1e - 16$.

In the absence of any other information, we start the iterations with the initial vector $\boldsymbol{p}_0 = (0.5, 0.5, 0.5)^\top$, which values are as neutral as possible. We use the same initial values $B_0 = 6125, H_0 = 5362$.

We recommend using $I_{\mathrm{B}} = I_{\mathrm{H}} = 10$ equidistantly distributed observation times, see Table 1.

Table 1. IP1 for 1975 data

Parameter	p_0^i	p^i	$\breve{p}^i$	$\lvert p^i - \breve{p}^i \rvert$	$\frac{\lvert p^i - \breve{p}^i \rvert}{p^i}$	r^i
μ_B	0.50	0.575	0.573	2.000e−3	3.478e−3	3.50e−10
ϕ	0.50	0.571	0.570	9.998e−4	1.751e−3	2.80e−10
μ_H	0.50	0.302	0.301	9.999e−4	3.311e−3	3.10e−10

The parameters are accurately reconstructed, where the absolute error is around 1e−3.

Let us test the data for 1976. Now $\boldsymbol{p} = (\mu_B, \phi, \mu_H)^\top = (0.583, 0.571, 0.246)^\top$. We again use $\boldsymbol{p}_0 = (0.5, 0.5, 0.5)^\top$ and $B_0 = 5982, H_0 = 5362$. The results are displayed on Table 2.

Table 2. IP1 for 1976 data

Parameter	p_0^i	p^i	$\check{p}^i$	$\lvert p^i - \check{p}^i \rvert$	$\frac{\lvert p^i - \check{p}^i \rvert}{p^i}$	r^i
μ_B	0.50	0.583	0.582	1.000e−3	1.716e−3	3.20e−10
ϕ	0.50	0.571	0.570	1.000e−3	1.752e−3	2.80e−10
μ_H	0.50	0.246	0.245	9.999e−4	4.065e−3	3.10e−10

The absolute errors again are small, implying a good fit and precisely recovered parameters.

Both tables demonstrate the effectiveness of the optimization approach in recovering the parameters with high accuracy, ensuring that the model can predict honeybee (and brood) population dynamics effectively using real data.

5 Discussion

5.1 Results

The results of the computational simulations demonstrate that the proposed parameter optimization approach is effective in accurately recovering key model parameters for both the 1975 and 1976 data sets. The absolute errors, consistently around 1e−3, are sufficiently small to suggest a precise fit between the model's predictions and the actual data. This level of accuracy is critical, as it ensures that the model is capable of capturing the diffusion dynamics of honeybee and brood populations, even under varying conditions.

Accurate parameter recovery is essential for generating reliable simulations, which, in turn, allows for more accurate forecasting and assessment of population dynamics. In ecological models such as (1), small errors in parameter estimates can lead to significant deviations in long-term predictions. Thus, the precise reconstruction of parameters like brood death rates and adult bee transition rates allows for better simulation of potential scenarios and enables researchers and beekeepers to identify harmful processes in their early stages. This is particularly important in honeybee population studies, where early detection of colony collapse factors can significantly influence conservation and intervention strategies.

Moreover, the ability to recover parameters with such precision implies that the optimization method is robust, even when using real-world data, which often contains noise and variability. This further underscores the model's utility for practical applications, offering a tool that can be used to track colony health and stability over time, thereby helping to mitigate the risks, including those associated with population decline.

5.2 Choice of a Numerical Method

The adjoint equations optimization approach offers several advantages in solving inverse problems, particularly in models involving complex systems like honeybee population dynamics. One of its main strengths is the computational efficiency it provides.

By leveraging the adjoint system, the method allows for the gradient of the error functional to be computed with respect to all parameters in a single backward sweep, irrespective of the number of parameters involved. This reduces the computational cost significantly compared to traditional methods, especially in high-dimensional problems where multiple parameters must be optimized simultaneously. Additionally, the adjoint approach is well-suited for problems where the number of parameters is large, as it does not require evaluating the objective function for each individual parameter, making it scalable and efficient for large-scale applications.

However, the adjoint method also comes with certain challenges. One potential disadvantage is the need to derive and solve the adjoint system, which can be analytically and numerically complicated, particularly for nonlinear or highly sophisticated models. The complexity of solving the adjoint problem can introduce difficulties in implementation, especially when compared to more direct numerical approaches. Moreover, the convergence of the adjoint method depends heavily on the accuracy of the adjoint system solution, and errors in the adjoint problem can propagate, leading to inaccuracies in the parameter estimates.

When compared to other optimization methods, such as the trust-region approach, which directly minimizes the error functional, the adjoint method offers faster gradient computation but may be less robust in handling non-convex or noisy error landscapes. The trust-region method, for example, adjusts the step size dynamically to stay within a region where the functional approximation remains valid, making it more stable in certain cases, especially when dealing with complex objective functions. However, this comes at the cost of higher computational expenses, as the trust-region method often requires multiple evaluations of the error functional at different points within the region to refine the step size, leading to slower convergence in large-scale problems.

6 Conclusion

In this study, we developed and tested a mathematical model to understand the influence of brood death rates on honeybee population dynamics. By formulating and solving inverse problems, we aimed to recover critical parameters of the model that are otherwise difficult to measure directly. The accuracy of the parameter reconstruction was demonstrated through computational experiments using historical data from 1975 and 1976.

The described computational algorithm offers significant benefits to beekeepers. With an accurate parameter reconstruction algorithm, reliable simulations of honeybee population dynamics can be performed. These simulations enable the detection of harmful processes in their initial phases, allowing for timely interventions that can prevent colony collapse. By understanding how different parameters affect the system, beekeepers can manage their colonies more effectively, mitigating the severity of adverse conditions or potentially eradicating the effects of disorders on the hives.

Additionally, the model provides valuable insights into the ecological and biological processes governing honeybee populations. By incorporating brood mortality into the model, we offer a more comprehensive understanding of the factors influencing colony health and survival. This holistic approach is crucial for developing effective strategies to protect honeybee populations, which are vital for pollination and the broader ecosystem.

In a future work we plan to extend this model to a delayed ordinary differential equations system describing the effect of insecticides over brood death rate and its influence over honeybee population dynamics as well as to study a similar optimization problem for this new model.

Future work may also focus on enhancing the current algorithm and testing it with more sophisticated dynamics models. Improvements to the model could include incorporating additional variables and interactions that influence honeybee population dynamics, such as environmental factors, disease transmission, and genetic variability. These enhancements will provide a more precise description of the dynamics, enabling deeper analysis and more accurate predictions.

Further research might also explore advanced optimization techniques to improve the efficiency and robustness of the parameter reconstruction process. By refining the algorithm, we aim to achieve even greater accuracy in parameter estimation, which will support beekeepers in making data-driven decisions for colony management.

Moreover, the development of user-friendly software tools based on this model can make the benefits accessible to a broader community of beekeepers. These tools can provide real-time monitoring and simulation capabilities, helping beekeepers detect and respond to potential threats promptly.

Acknowledgement. This study was funded by the Bulgarian National Science Fund under Project KP-06-PN 46–7 "Design and research of fundamental technologies and methods for precision apiculture".

Data Availibility Statement. Data and implemented software are available from the authors upon reasonable request.

References

1. Atanasov, A.Z., Georgiev, S.G., Vulkov, L.G.: Reconstruction analysis of honeybee colony collapse disorder modeling. Optim. Eng. **22**(4), 2481–2503 (2021). https://doi.org/10.1007/s11081-021-09678-0
2. Becher, M.A., Osborne, J.L., Thorbek, P., Kennedy, P.J., Grimm, V.: Review: towards a systems approach for understanding honeybee decline: a stocktaking and synthesis of existing models. J. Appl. Ecol. **50**, 868–880 (2013)
3. Betti, M.I., Wahl, L.M., Zamir, M.: Effects of infection on honey bee population dynamics: a model. PLoS ONE **9**(10), e110237 (2014)
4. Camazine, S., Sneyd, J., Jenkins, M.J., Murray, J.D.: A mathematical model of self-organized pattern formation on the combs of honeybee colonies. J. Theor. Biol. **147**(4), 553–571 (1990)
5. Cascante, J.E., et al.: Influence of brood deaths on Honey Bee population dynamics and the potential impact of insecticides. IBIO 4299 International Research Experience for Students IRES (2017)
6. Cox-Foster, D.L., et al.: A metagenomic survey of microbes in honey bee colony collapse disorder. Sci. **318**, 283–287 (2007)

7. Dornberger, L., et al.: Death of the bees: a mathematical model of colony collapse disorder. technical report 2012-12, Mathematics Preprint Series, University of Texas at Arlington Mathematics Department (2012)
8. Harris, J.L.: A Population Model and Its Application to The Study of Honey Bee Colonies. Master Thesis. The University of Manitoba, Department of Entomology (1980)
9. Eberl, H.J., Frederick, M.R., Kevan, P.G.: Importance of brood maintenance terms in simple models of the honeybeevarroa destructor-acute bee paralysis virus complex. Electron. J. Diff. Eqns. **19**, 85–98 (2010)
10. Kang, Y., Rodriguez-Rodriguez, M., Evilsizor, S.: Ecological and evolutionary dynamics of two-stage models of social insects with egg cannibalism. J. Math. Anal. Appl. **430**(1), 324–353 (2015)
11. Aronstein, K.A., Saldivar, E., Vega, R., Westmiller, S., Douglas, A.E.: How varroa parasitism affects the immunological and nutritional status of the honey bee, apis mellifera. Insects **3**(3), 601–615 (2012)
12. Khoury, D.S., Myerscough, M.R., Barron, A.B.: A quantitative model of honey bee colony population dynamics. PLoS ONE **6**(4), e18491 (2011)
13. Khoury, D.S., Barron, A.B., Meyerscough, M.R.: Modelling food and population dynamics honey bee colonies. PLoS ONE **8**(5), e0059084 (2013)
14. Kribs-Zaleta, C.M., Mitchell, C.: Modeling colony collapse disorder in honeybees as a contagion. Math. Biol. Eng. **11**(6), 1275–1294 (2014)
15. Marchuk, G.I.: Adjoint Equations and Analysis of Complex Systems. Kluwer, Dordrecht (1995)
16. Marchuk, G.I., Agoshkov, V.I., Shutyaev, V.P.: Adjoint Equations and Perturbation Algorithms in Nonlinear Problems. CRC Press, Boca Raton (1996)
17. MAAREC. The colony and its organization. https://canr.udel.edu/maarec/honey-bee-biology/the-colony-and-its-organization/ (2024). Accessed 11 Apr 2024
18. Mid-Atlantic Apiculture Research and Extension Consortium, Basic bee biology for beekeepers; Fact Sheet, MAAREC Publication 1.4 March 2004. https://bee-health.extension.org/basic-bee-biology-for-beekeepers/. Accessed 11 Apr 2024
19. Oldroyd, B.P.: What's killing american honey bees? PLoS Biol. **5**(6), e168 1195–1199 (2007)
20. Perry, C.J., Søvik, E., Myerscough, M.R., Barron, A.B.: Rapid behavioral maturation accelerates failure of stressed honey bee colonies. Proc. Natl. Acad. Sci. **112**(11), 3427–3432 (2015)
21. Russel, S., Barron, A.B., Harris, D.: Dynamics modelling of honeybee (Apis mellifera) colony growth and failure. Ecolog. Model. **265**, 138–169 (2013)
22. Van der Zee, R., et al.: Managed honey bee colony losses in Canada, China, Europe, Israel and Turkey for the winters of 2008–2009 and 2009–2010. J. Appl. Res. **51**(1), 100–114 (2012)
23. vanEngelsdorp, D., et al.: Colony collapse disorder: a descriptive study. PLoS ONE **4**(8), e6481 (2009)
24. Winston, W.L.: The Biology of the Honey Bee. Harvard University Press, Cambridge Mass (1991)

The Study of the Influence of Green House Pollutants on the Adiabatic Constant of the Atmospheric Air

Mihai Petrov[1], Velichka Traneva[1(✉)], Stoyan Tranev[2], Venelin Todorov[3,5], and Slavi G. Georgiev[3,4]

[1] Department of Mathematics, Informatics and Physics, 1 Prof. Yakim Yakimov Blvd., 8010 Burgas, Bulgaria
mihpetrov@abv.bg, veleka13@gmail.com

[2] Department of Industrial Technology and Management, 1 Prof. Yakim Yakimov Blvd., 8010 Burgas, Bulgaria
tranev@abv.bg

[3] Institute of Mathematics and Informatics, Bulgarian Academy of Sciences, Acad. G. Bonchev Street, Bl. 8, 1113 Sofia, Bulgaria
sggeorgiev@math.bas.bg

[4] Department of Applied Mathematics and Statistics, University of Ruse, 8 Studentska Street, 7004 Ruse, Bulgaria

[5] Institute of Information and Communication Technologies, Bulgarian Academy of Sciences, 25 A Acad. Georgi Bonchev Street, 1113 Sofia, Bulgaria
venelin@iict.bas.bg

Abstract. The state of the atmosphere is significantly affected by greenhouse gaseous pollutants. The respective thermodynamic parameters that describe the state of the atmosphere are its pressure, temperature and volume. Anthropogenic gases with a greenhouse effect influence these parameters and especially the average global temperature of the atmosphere. As a result of comparing the temperature records and the concentrations of pollutants such as carbon dioxide, which has the highest ratio from all green house gases, it is concluded that there is a quantitative expression based on the laws of thermodynamics between the temperature values recorded over time and the concentration values. As a result of the application of the laws of thermodynamics of the atmospheric state, the quantitative expressions are obtained that allow calculating the adiabatic constant of atmospheric air. By the application of the temperature gradient as a function of the altitude, the following fact is found that the absolute value of the gradient increases over time, a fact that leads to a slow instability of the atmosphere with values of the adiabatic constant that decrease slowly over time. As a manner of the validation of the suggested method, the numerical value of the universal gas constant is re-calculated. Another important obtained result is about specifical thermal capacity of atmospheric air. This value is decresaing slowly over time together with the decreasing of the value of adiabatical constant. The slowly decreasing of both parameters leads to the increasing of the speed of winds which are more frequent last time. The importance of the suggested method of the paper consists in the fact that the parameters that are contained in the expressions allows to calculate the speed of winds and hurricanes which are increasing over time.

D. Simian and L. F. Stoica (Eds.): MDIS 2024, CCIS 2486, pp. 183–199, 2025.
https://doi.org/10.1007/978-3-031-87386-7_13

Keywords: Green House Pollutants · Temperature · Adiabatic Constant · Temperature Gradient · Winds and Hurricanes

1 Introduction

The increasing concentration of anthropogenic greenhouse gases in the atmosphere and biosphere has a negative impact on flora, fauna, and human health. Recent extreme temperature fluctuations, characterized by rapid variations, are a result of atmospheric instability. This instability can be quantitatively described using thermodynamic laws, which allow for the prediction of atmospheric behavior and variations in key parameters.

A fundamental aspect of these extreme temperature changes over short time intervals is the adiabatic constant of the atmosphere. Temperature gradients as a function of altitude vary due to changes in atmospheric composition, influenced by pollutant concentrations over time.

This study aims to develop a quantitative theoretical framework based on thermodynamic principles to calculate essential atmospheric parameters and predict the overall state of the atmosphere in conditions of excessive greenhouse gas accumulation. The primary objective is to analyze the dependence of the atmospheric adiabatic constant and specific heat capacity on pollutant concentrations.

2 Material and Methods

The suggested method of this research with recent scientific novelty consists in the development of new expressions for the description of the atmospheric state. In fact, this is the purpose of the work to establish an explanation of the phenomena that take place for the dynamics of the atmospheric state by the application of thermodynamical description.

The Earth's atmosphere is divided into five main atmospheric layers: tropo-sphere, stratosphere, mesosphere, thermosphere, exosphere, and the boundaries between them are not strictly defined. The first and closest atmospheric layer to the Earth's surface is called the troposphere [1]. The temperature of the tropo-sphere is highest near the Earth's surface and decreases with altitude. The average temperature gradient of the troposphere is 6.5 °C at 1000 m above sea level. The adiabatic temperature gradient: $-\frac{dT}{dh}$ (h is the altitude) [2–4]. The rate at which the temperature decreases with height is called the adiabatic temperature gradient. The adiabatic equation for the ideal gas is valid for dry air [5–7]:

$$P(h) \cdot [T(h)]^{-\frac{\gamma}{\gamma-1}} = const, \tag{1}$$

where γ is the adiabatic constant of air ($\gamma \approx 7/5$). The combination with the equation of pressure gives the expression for the dry adiabatic temperature gradient as follow [8]:

$$\frac{dT}{dh} = -\frac{M \cdot g}{R} \cdot \frac{\gamma - 1}{\gamma} = -9,8^0 C/km. \tag{2}$$

Atmospheric pressure is maximal above the sea level and decreases with altitude. The application of the expression (1) with the logarithmatization operation gives the following result:

$$\frac{\gamma}{\gamma - 1} = \frac{\ln P - \ln Const}{\ln T}. \tag{3}$$

The right side of the expression (3) can be signed simply by *Z*.
Then

$$\gamma = \frac{Z}{Z - 1}. \tag{4}$$

The aim of the paper is the obtaining of the expression of the variation of adiabatical constant of gases $\Delta\gamma$. The derivative of the expression (3) gives:

$$\gamma^{/} = \Delta\gamma = \frac{Z^{/}}{Z - 1} + Z \cdot (-1) \cdot (Z - 1)^{-2} \cdot Z^{/} = \frac{-Z^{/}}{(Z - 1)^2}. \tag{5}$$

The derivative of the value *Z* gives:

$$Z^{/} = \frac{(\ln P)^{/} - (\ln P - \ln Const) \cdot (\ln T)^{/}}{(\ln T)^2}. \tag{6}$$

$$Z^{/} = \frac{\frac{\Delta P}{P} - (\ln P - \ln Const) \cdot \frac{\Delta T}{T}}{(\ln T)^2}. \tag{7}$$

The substitution of (1) into (7) gives:

$$Z^{/} = \frac{\frac{\Delta P}{P} - \left(\frac{\gamma}{\gamma - 1} \cdot \ln T\right) \cdot \frac{\Delta T}{T}}{(\ln T)^2}. \tag{8}$$

The atmosphere can be approximated by the model of ideal gas and the equation of ideal gas is written as:

$$P = C \cdot R \cdot T, \tag{9}$$

where *P* is the pressure of atmosphere; *C*- the concentration of the atmospheric air; *T*-temperature of atmosphere; *R*- universal constant of gases ***(R = 8,31 (J/(mol.K))***.

The variation of the pressure of atmosphere due to of the variation of its concentration ΔC which is the result of accumulation of green house gases is written as:

$$\Delta P = R \cdot (\Delta T \cdot C + \Delta C \cdot T). \tag{10}$$

The substitution of (11) into (5) gives:

$$Z^{/} = \frac{\gamma \cdot \frac{\Delta C}{C} - \frac{\Delta T}{T} \cdot \frac{\gamma}{\gamma - 1} \cdot \ln T}{(\ln T)^2}. \tag{11}$$

The substitution of (11) into (5) gives:

$$\Delta\gamma = \frac{\frac{\Delta T}{T} \cdot \frac{\gamma}{\gamma-1} \cdot \ln T - \gamma \cdot \frac{\Delta C}{C}}{(\ln T)^2 \cdot \left(\frac{\ln P - \ln Const}{\ln T} - 1\right)^2}. \tag{12}$$

The substitution of (10) into (9) gives the following expression of the temperature variation:

$$\Delta T = (\gamma - 1) \cdot \frac{\Delta C \cdot T}{C}. \tag{13}$$

The substitution of the expression (13) into (12) gives:

$$\Delta\gamma = \frac{\frac{\Delta T}{T} \cdot \frac{\gamma}{\gamma-1} \cdot \ln T - \gamma \cdot \frac{\Delta C}{C}}{\left(\frac{\gamma}{\gamma-1} \cdot \ln T - \ln T\right)^2} = \frac{(\gamma - 1) \cdot \gamma \cdot \Delta T \cdot (\ln T - 1)}{T \cdot (\ln T)^2}. \tag{14}$$

$$\gamma - 1 = \frac{T \cdot (\ln T)^2 \cdot \Delta\gamma}{\gamma \cdot \Delta T \cdot (\ln T - 1)}. \tag{15}$$

The substitution of (15) into (2) gives the following:

$$\frac{dT}{dh} = -\frac{M \cdot g}{R} \cdot \frac{T \cdot (\ln T)^2 \cdot \Delta\gamma}{\gamma^2 \cdot \Delta T \cdot (\ln T - 1)}. \tag{16}$$

$$\Delta\gamma = -\frac{\gamma^2 \cdot R \cdot (\ln T - 1) \cdot \frac{dT}{dh} \cdot \Delta T}{T \cdot (\ln T)^2 \cdot M \cdot g}. \tag{17}$$

According to the expression (17) we expect that the adiabatic constant of atmospheric air is decreasing slowly over time with the increasing of the atmospheric temperature (or with the increasing of green house pollutants).

The most important peculiarity of the expression (17) is that when the variation of temperature is null ($\Delta T = 0$), then there is no any change of adiabatical constant ($\Delta\gamma = 0$).

3 Discussion of Results

In order to calculate the variation of the adiabatic constant of atmospheric air is necessary firstly to calculate the molar concentration of atmospheric air.

The density of air at sea level is about 1,2 kg/m^3. The molar concentration of the atmospheric air is calculated as:

$$C = \frac{\rho}{M} = \frac{1,2(kg/m^3)}{0,029(kg/mol)} = 41,37(mol/m^3) = 0,04137(mol/l). \tag{18}$$

The calculated value of molar concentration of air is used to appreciate the change of the temperature as the result of the increasing of greenhouse gases by the value ΔC.

In general the average temperature of the atmospheric air is in-creasing almost linearly last 50 years [9].

The opinions about the reason of the increasing of the temperatures are various, but the most observation is on the focusing of the increasing of the concentration of carbon dioxide that is one major greenhouse gas and the carbon dioxide has 75% effect from all others green house gases [10].

Beside the registrations of the temperatures, there is also the registration of the variations with time of the concentrations of carbon dioxide [11].

The systematization of the data about the variations of the temperatures [9] and concentrations [11] with the further calculations are represented on the Table 1.

The usually concentrations that are represented on the Table 1 as ppm values can be transformed into mg/m^3 by the application of the method given in [12]. So, the respective variation of ppm is:

$$\Delta C_{ppm} = \frac{24,4(l/mol) \cdot \Delta C[mg/m^3]}{M[g/mol]} . \tag{19}$$

The variation of the concentrations of carbon dioxide $\Delta C[mg/m^3]$ in the atmosphere is:

$$\Delta C[mg/m^3] = \frac{M[g/mol] \cdot \Delta C_{ppm}}{24,4} . \tag{20}$$

The conversion from *ppm* into *mol/l* is performed as:

$$\Delta C[mol/l] = \frac{\Delta C_{ppm}}{1000 \cdot M[g/mol]} . \tag{21}$$

The calculation of molar concentrations of the excess of carbon dioxide is represented on the Table 1.

In order to validate the suggested method of this study, it is necessary numerically checking of one of the parameters that are contained in the expressions. Regarding the expression (17) it can be concluded that it could be checked if the value of the universal gas constant R satisfies the tabular numerical values. For this purpose, the graph of the dependence $\Delta\gamma = f(\Delta T)$ based on expression (14) could be plotted. The calculations according to the expression (14) could be used for the expression (17) in order to check the numerical value of the universal gas constant according to the proportionality coefficient. The proportionality coefficient of the expression (17) is:

$$K = tg\alpha = \frac{\gamma^2 \cdot R \cdot (\ln T - 1) \cdot \frac{dT}{dh}}{T \cdot (\ln T)^2 \cdot M \cdot g} . \tag{22}$$

The last column of the Table 1 contains the calculated values of the modules of $|\Delta\gamma|$. The graphic of the dependence of $|\Delta\gamma| = f(\Delta T)$ is represented on the Fig. 1.

The correlation calculations of the Fig. 1 shows that the proportional coefficient K is $K = 28,273 \cdot 10^{-5}$. The value of temperature T is considered the average global temperature 15 0C (288 K). The adiabatic constant $\gamma = 1,4$ for atmospheric air [13]. The

Table 1. Table variation of temperature as the function of the increasing of the concentration of CO_2 [10].

N°	Year [9, 11]	ΔT; °C [9]	C (CO_2); Ppm [11]	ΔC (CO_2);ppm	ΔC (CO_2); mol/l	ΔC (CO_2); mg/m^3	$\|\Delta\gamma\| \times 10^{-5}$
0	1980	0	338	0	0	0	0
1	1981	0,03	340	2	4,54545E-05	3,61	8,48187E-06
2	1982	0,05	341	3	6,81818E-05	5,41	1,41364E-05
3	1983	0,07	343	5	0,000113636	9,01	1,9791E-05
4	1984	0,09	345	7	0,000159091	12,63	2,54456E-05
5	1985	0,11	347	9	0,000204545	16,23	3,11002E-05
6	1986	0,13	348	10	0,000227273	18,03	3,67548E-05
7	1987	0,15	349	11	0,00025	19,84	4,24093E-05
8	1988	0,18	350	12	0,000272727	21,64	5,08912E-05
9	1989	0,19	350.5	12,5	0,000284091	22,54	5,37185E-05
10	1990	0,22	351	13	0,000295455	23,44	6,22004E-05
11	1991	0,26	352	14	0,000318182	25,25	7,35095E-05
12	1992	0,32	353	15	0,000340909	27,05	9,04732E-05
13	1993	0,34	355	17	0,000386364	30,66	9,61278E-05
14	1994	0,36	357	19	0,000431818	34,26	0,000101782
15	1995	0,37	359	21	0,000477273	37,87	0,00010461
16	1996	0,39	360	22	0,0005	39,67	0,000110264
17	1997	0,41	361	23	0,000522727	41,48	0,000115919
18	1998	0,42	362	24	0,000545455	43,28	0,000118746
19	1999	0,43	363	25	0,000568182	45,08	0,000121573
20	2000	0,45	365	27	0,000613636	48,69	0,000127228
21	2001	0,46	369	31	0,000704545	55,90	0,000130055
22	2002	0,47	373	35	0,000795455	63,12	0,000132883
23	2003	0,5	375	37	0,000840909	66,72	0,000141364
24	2004	0,54	376	38	0,000863636	68,52	0,000152674
25	2005	0,58	378	40	0,000909091	72,13	0,000163983
26	2006	0,59	379	41	0,000931818	73,93	0,00016681
27	2007	0,6	380	42	0,000954545	75,74	0,000169637
28	2008	0,61	381	43	0,000977273	77,54	0,000172465
29	2009	0,62	382	44	0,001	79,34	0,000175292

(continued)

Table 1. (*continued*)

N°	Year [9, 11]	ΔT; °C [9]	C (CO_2); Ppm [11]	ΔC (CO_2);ppm	ΔC (CO_2); mol/l	ΔC (CO_2); mg/m^3	$\|\Delta\gamma\|$ x 10^{-5}
30	2010	0,63	383	45	0,001022727	81,15	0,000178119
31	2011	0,65	385	47	0,001068182	84,75	0,000183774
32	2012	0,68	393	55	0,00125	99,18	0,000192256
33	2013	0,7	398	60	0,001363636	108,20	0,00019791
34	2014	0,74	401	63	0,001431818	113,61	0,000209219
35	2015	0,78	402	64	0,001454545	115,41	0,000220529
36	2016	0,8	406	68	0,001545455	122,63	0,000226183
37	2017	0,81	410	72	0,001636364	129,84	0,00022901
38	2018	0,82	412	74	0,001681818	133,44	0,000231838
39	2019	0,83	413	75	0,001704545	135,25	0,000234665
40	2020	0,85	415	77	0,00175	138,85	0,00024032

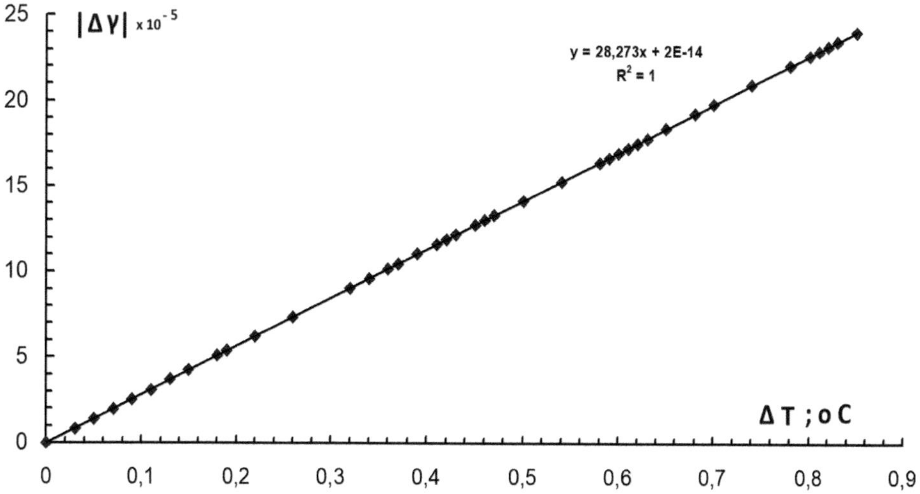

Fig. 1. The variation of adiabatic constant of atmospheric air as the function of the variation of atmospheric temperature.

values for the calculations both for expression (14) and (17) are considered as 288 K and $\gamma = 1{,}4$ nevertheless that are slowly changeable with time.

The universal constant of gases then can be calculated as:

$$R = \frac{T \cdot (\ln T)^2 \cdot M \cdot g \cdot K}{\gamma^2 . (\ln T - 1) \cdot \frac{dT}{dh}} = \frac{288 \cdot (\ln 288)^2 \cdot 0,029 \cdot 9,83 \cdot 28,273 \cdot 10^{-5}}{1,4^2 \cdot (\ln 288 - 1) \cdot 0,0098} = 8,3111 (J/(mol \cdot K)). \quad (23)$$

The tabular value is $R = 8{,}314\ (J/(mol{\cdot}K))$. It is almost coincident with maximum possible error of exactity:

$$\varepsilon\% = \frac{|8,314 - 8,3111|}{8,314}.100\% = \frac{0,0029 \cdot 100\%}{8,314} = 0,03\%\,. \tag{24}$$

In such a way the suggested method for the description of the state of atmosphere by classical laws of thermodynamics is valid and applicable.

The functional dependence of Fig. 1 gives possibility to write the empirical dependence of the variation of adiabatical constant as the function of the variation of atmospheric temperature as follow:

$$\Delta\gamma = -28,273 \cdot 10^{-5} \cdot \Delta T\left({}^{0}C\right). \tag{25}$$

The value $C = 0{,}04137\ (mol/l)$ for the whole atmospheric air.

The expression (13) gives the possibility to represent the graphical dependence of the variation of the temperature of atmosphere as the function of the variation of concentrations of green house gases $\Delta T = f(\Delta C)$. (Fig. 2).

Considering the expression from (13) we obtain:

$$\frac{\Delta T}{T} = (\gamma - 1) \cdot \frac{\Delta C}{C}\,. \tag{26}$$

The proportional coefficient is

$$477,26 = \frac{T}{C} \cdot (\gamma - 1). \tag{27}$$

Then, the calculation of adiabatic constant is performed as:

$$\gamma - 1 = \frac{477,26 \cdot 0,04137}{288} = 0,068 \text{ hence } \gamma = 1,068\,. \tag{28}$$

Regarding Fig. 2 it would be possible to write the empirical dependence about the variation of the atmospheric temperature as the function of the variation of molar concentration of green house gases:

$$\Delta T({}^{0}C) = 477,26 \cdot \Delta C(mol/l)\,. \tag{29}$$

The similar dependence of the variation of adiabatical constant of atmospheric air as the function of the variation of the concentration can be represented on the Fig. 3.

$$|\Delta\gamma| = f(\Delta C) \tag{30}$$

The expression (14) can be re-written by the application of expression (13):

$$\Delta\gamma = \frac{(\gamma - 1) \cdot \gamma \cdot \Delta T \cdot (\ln T - 1)}{T \cdot (\ln T)^2} = \frac{(\gamma - 1)^2 \cdot \gamma \cdot \Delta C \cdot (\ln T - 1)}{C \cdot (\ln T)^2}\,. \tag{31}$$

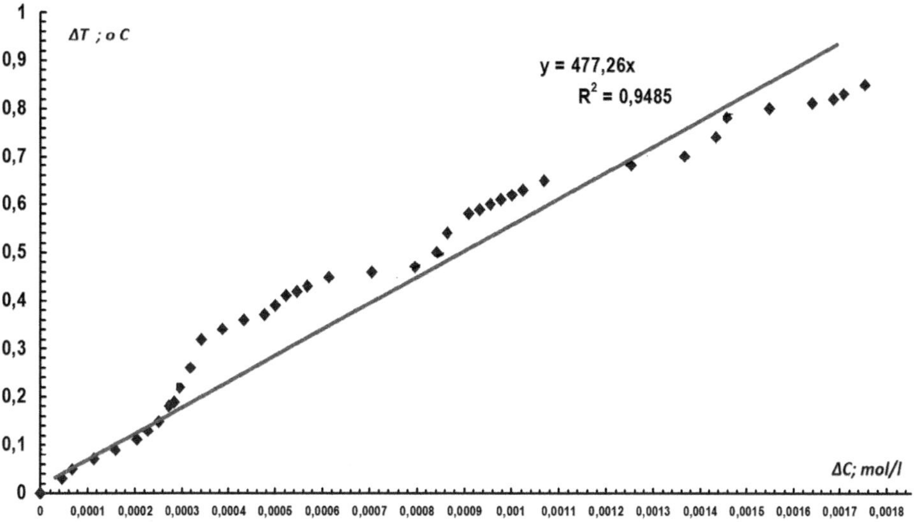

Fig. 2. The variation of the temperatures of atmosphere ΔT as the function of accumulation of green house gases ΔC.

The expression (17) can be also re-written by the application of the expression (13):

$$|\Delta\gamma| = \frac{\gamma^2 \cdot R \cdot (\ln T - 1) \cdot \frac{dT}{dh} \cdot \Delta T}{T \cdot (\ln T)^2 \cdot M \cdot g} = \frac{\gamma^2 \cdot (\gamma - 1) \cdot R \cdot (\ln T - 1) \cdot \frac{dT}{dh} \cdot \Delta C}{C \cdot M \cdot g \cdot (\ln T)^2}. \tag{32}$$

Comparing the right sides of both above expressions, it would give possibility to re-calculate the value of adiabatical constant γ of atmospheric air:

$$\frac{(\gamma - 1)^2 \cdot \gamma \cdot \Delta C \cdot (\ln T - 1)}{C \cdot (\ln T)^2} = \frac{\gamma^2 \cdot (\gamma - 1) \cdot R \cdot (\ln T - 1) \cdot \frac{dT}{dh}}{C \cdot M \cdot g \cdot (\ln T)^2} \tag{33}$$

hence

$$\gamma = \frac{C.M.g}{C.M.g - R.\frac{dT}{dh}}. \tag{34}$$

$$\gamma = \frac{C \cdot M \cdot g}{C \cdot M \cdot g - R \cdot \frac{dT}{dh}} = \frac{0,04137 \cdot 10^3 (mol/m^3) \cdot 29 \cdot 10^{-3} (kg/mol) \cdot 9,83}{0,04137 \cdot 10^3 \cdot 9,83 - 8,31 \cdot 0,0098} = 1,007. \tag{35}$$

Regarding the Fig. 3, the empirical dependence of adiabatic constant as the function of the variation of molar concentration of pollutants is written as:

$$\Delta\gamma = -0,13494 \cdot \Delta C(mol/l). \tag{36}$$

The dependence of adiabatic constant as the function of interval of time is represented on the Fig. 4. The regression analysis of curve shows in the linear approach that it is

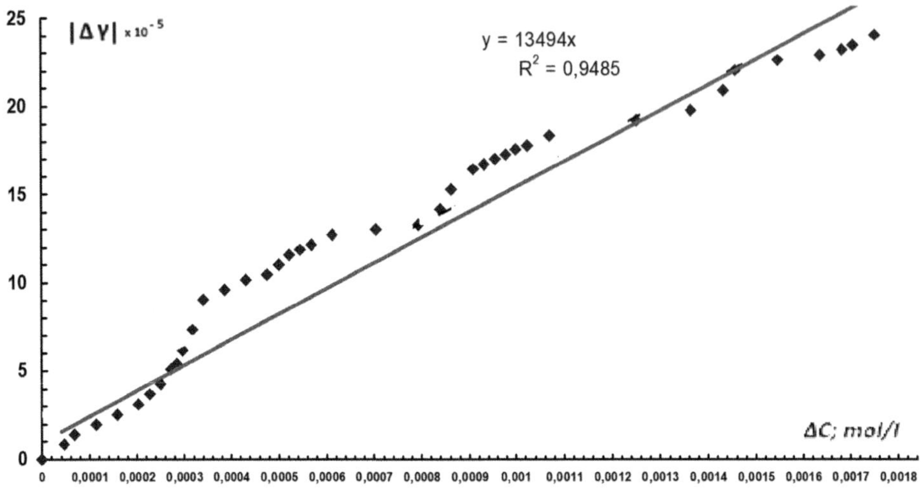

Fig. 3. The dependence of the variation of adiabatic constant as the function of the variation of the concentrations of green house gases.

possible to estimate the interval of time when the atmospheric air could reach the limit value of adiabatic constant almost equal to 1. Then:

$$1,4 - 0,0001 \cdot \Delta t = 1 \text{ hence } \Delta t = \frac{0,4}{0,0001} \leq 4000\,years. \tag{37}$$

The analysis of the data of adiabatic constant of air leads to the conclusion that the extreme limit value of $\gamma = 1$ can be reached by the linear approximation within the maximal limit interval of time of the order 4000 years with the consideration that the state of anthropogenical activity is performed at the same rate similar to the recent rate.

The atmosphere gets more unstable when the concentration of green house pollutants is increasing. This fact can be checked by the expression (16). According to the expression (16) we can conclude that if the adiabatic constant of atmospheric air decreases (*$\Delta\gamma$ < 0),* then the temperature gradient is increasing gradually with time. It means that the atmospheric air cools more rapidly with altitude if the pollutants are increased. For the same altitude the atmospheric air cools more rapidly at higher concentrations of pollutants. But the higher values of temperature gradient give more unstable temperature. Another expression of the calculation of temperature gradient is:

$$\frac{dT}{dh} = -\frac{g}{C_p}. \tag{38}$$

where g is gravitational acceleration; C_p – specific heat capacity of atmospheric air at constant pressure [14]. The value of C_p is expected to decrease slowly with the increasing of $\frac{dT}{dh}$. The gradually decreasing of specific heat capacity over time leads to more rapid change of the values of temperatures of atmosphere. The atmosphere cools rapidly and warms rapidly at smaller values of C_p and in fact it is observed more frequently last time.

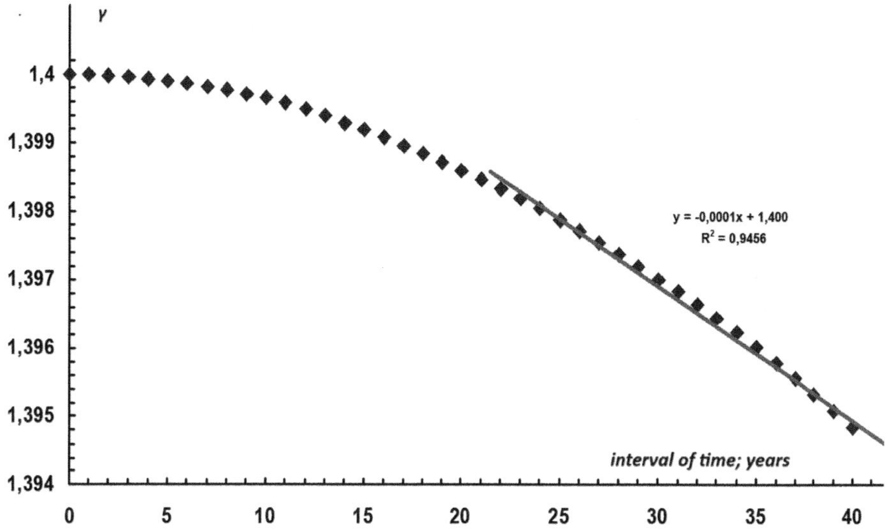

Fig. 4. The dependence of the adiabatic constant of atmospheric air as the function of time.

The difference of temperatures between various regions leads to the apparition of the winds. They are frequent last time. Higher frequency of the apparition of winds is explained by the decreasing of the specific heat capacity of atmospheric air [14]. The speed of winds is calculated by the formula:

$$u = \sqrt{2 \cdot g \cdot h \cdot \left(\frac{|T_2 - T_1|}{T_1}\right)}. \tag{39}$$

where h is the altitude with the corresponding smaller temperature. The value of T_1 is the temperature at the null level. The difference of temperatures leads to the apparition of winds. This difference $T_2 - T_1$ corresponds to the altitude h. The difference of temperatures $\Delta T_h = |T_2 - T_1|$ can be expressed by the temperature gradient from the expression (2):

$$dT = -\frac{M.g}{R}\left(\frac{\gamma - 1}{\gamma}\right)dh. \tag{40}$$

$$\Delta T_h = |T_2 - T_1| = \frac{dT}{dh} \cdot h. \tag{41}$$

Then, the expression of the speed of wind can be written as:

$$u = \sqrt{2 \cdot g \cdot h \cdot \frac{(dT/dh) \cdot h}{T_1}} = g \cdot h \cdot \sqrt{\frac{2}{C_p \cdot T_1}}. \tag{42}$$

Before to represent the data about the speed of winds, it is necessary to specify some calculations. First of all, it is necessary to calculate the temperature gradient as the function of the temperature of atmosphere. The expression (16) can be specified as follow:

$$\frac{dT}{dh} = -\frac{M.g}{R} \cdot \frac{T \cdot (\ln T)^2 \cdot \Delta\gamma}{\gamma^2 \cdot \Delta T \cdot (\ln T - 1)}. \tag{43}$$

Here, the temperature in fact is $T = 288 + \Delta T$ and the respective value of T is increasing gradually with time. The adiabatical constant $\gamma = 1{,}4\text{-}\Delta\gamma$ is decreasing gradually over time.

The respective dependence of temperature gradient as the function of the temperature of atmosphere is represented as follow on the Fig. 5. In order to appreciate the speed of winds it is necessary to calculate the specifical heat capacity of atmospheric air for the various values of adiabatic constant. The respective dependence is represented on the Fig. 6. The specific heat capacity is decreased with the decreasing of adiabatic constant [15].

The regression expression allows estimating by the linear segment of the Fig. 6 the value of specific heat capacity for the extreme case of $\gamma = 1$.

$$Y = 651,92 \cdot x + 89,459. \tag{44}$$

Then

$$x = 1 \; and \; C_p = 651,92 + 89,459 = 741,379 \; (J/(mol.K)). \tag{45}$$

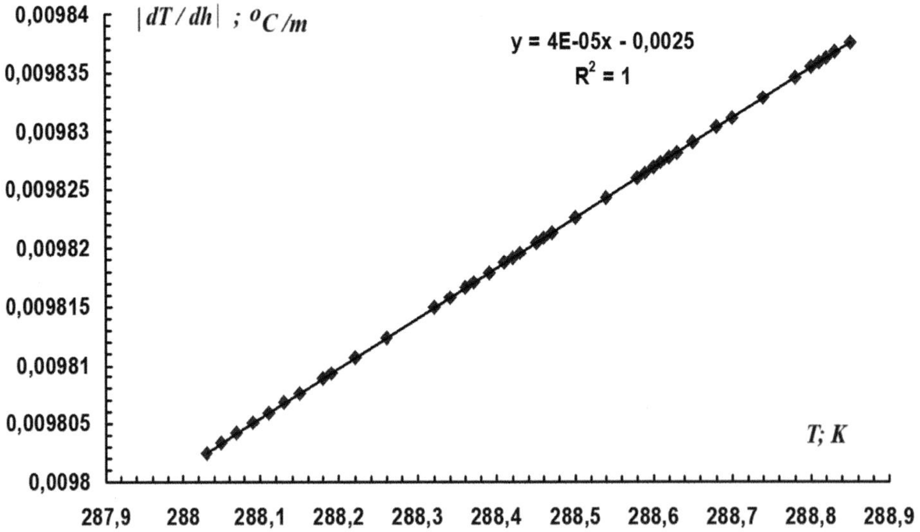

Fig. 5. The dependence of temperature gradient as the function of the temperature of atmosphere.

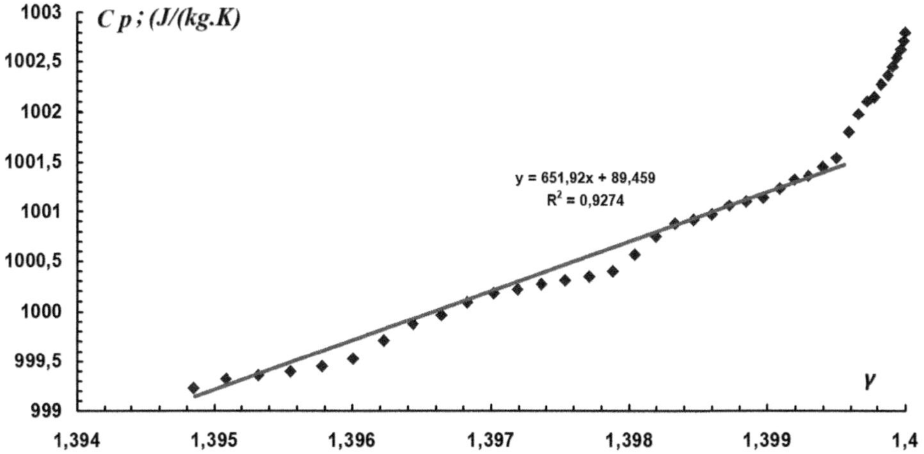

Fig. 6. The dependence of specific heat capacity as the function of adiabatic constant of air.

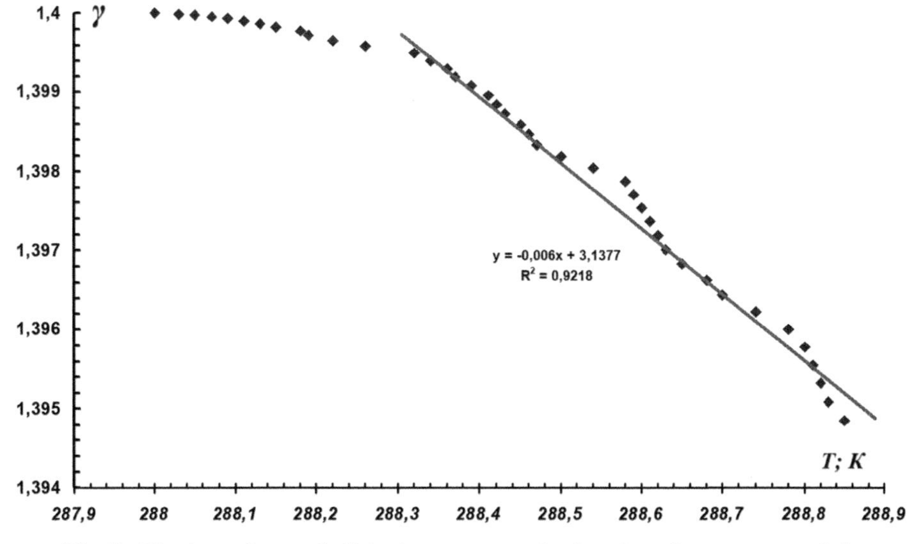

Fig. 7. The dependence of adiabatic constant as the function of temperature of air.

The dependence of adiabatic constant of air as the function of the temperature is represented on the Fig. 7.

The same situation of the decreasing of the adiabatic constant with the increasing of the temperature is represented on the paper [16].

The linear segment allows estimating of the temperature for the extreme case when the adiabatic constant is 1. The calculation gives the temperature of order 83 0C. This temperature of atmosphere hypothetically can be reached within 4000 years with the consideration that the anthropogenical activity with all pollutant emissions of the World's population has the same rate.

The dependence of specific heat capacity of atmospheric air as the function of the temperature of atmosphere is represented on the Fig. 8. Characteristically, is that the increasing of the temperature of atmosphere leads to the decreasing of specific heat capacity of air. The calculation of the values of the speed of wind as the function of the temperature of atmosphere at the ground level for the altitude of 100 m is represented on the Fig. 9. The calculations of the speed of wind is performed by the formula:

$$u = g \cdot h \cdot \sqrt{\frac{2}{C_p \cdot T_1}}. \tag{46}$$

The values of specific heat capacity are derived from the graph in Fig. 10, while atmospheric temperatures are represented as $T = 288 + \Delta T$.

Recent observations indicate that wind speeds are increasing at a much faster rate than they were 30 years ago [17]. The generalized data on wind speeds, as presented in Fig. 9, align with previous findings [17] and confirm a growing trend of more frequent and intense winds. These stronger winds have a negative and destructive impact on flora and fauna, further highlighting the consequences of atmospheric instability.

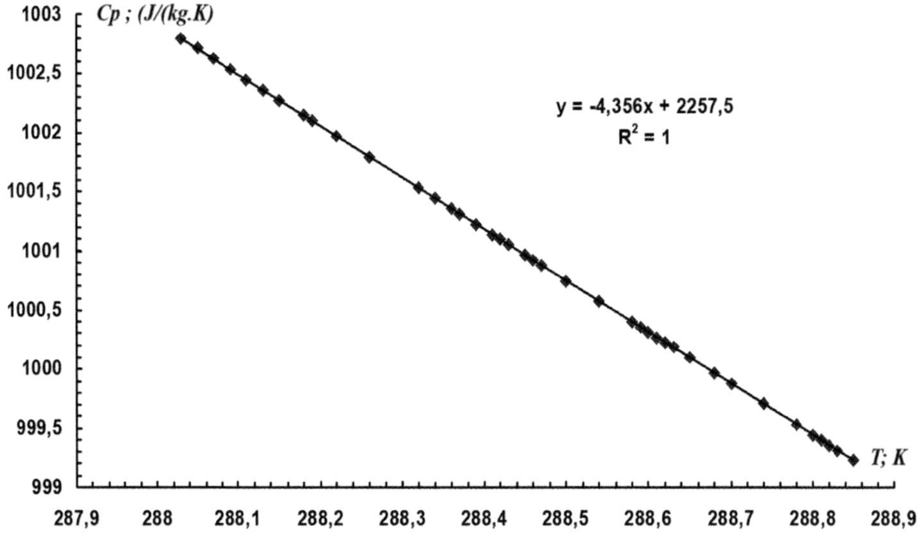

Fig. 8. The dependence of specific heat capacity of atmospheric air as the function of the temperature of atmosphere.

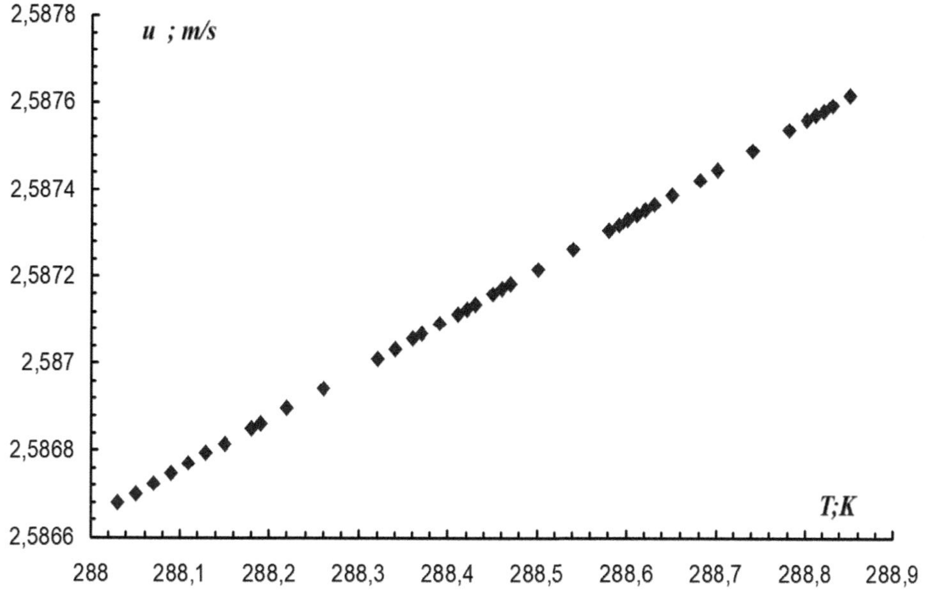

Fig. 9. The dependence of the speed of wind as the dependence of the temperature of atmosphere for the altitude $h = 100\ m$.

4 Conclusions

This study presents new results that provide a quantitative explanation of key aspects of atmospheric state dynamics. The scientific novelty lies in the application of the derivative of the adiabatic constant, allowing for a deeper understanding of atmospheric variations. The computed variations of the adiabatic constant were validated by recalculating the universal gas constant, which was found to match theoretical values within acceptable accuracy limits.

A significant finding of this study is that as global atmospheric temperatures rise due to increased greenhouse gas concentrations, the adiabatic constant of the atmosphere decreases. This decrease follows a linear trend with rising atmospheric temperatures. Furthermore, the graphical representation of temperature variations as a function of molar pollutant concentration enables precise determination of the adiabatic constant, confirming the validity of the proposed method.

Another crucial result is the long-term projection of atmospheric changes. When modeling the adiabatic constant as a function of time, it is estimated that, assuming the current pollution rate remains constant, the adiabatic constant would approach its extreme limit value of one within approximately 4,000 years. Additionally, when analyzing temperature gradients as a function of atmospheric temperature, it was found that the gradient modulus increases linearly. This means that at the same altitude, temperature decreases occur at a faster rate over time, leading to accelerated atmospheric cooling.

A further significant conclusion is that the decline in the adiabatic constant over time leads to a reduction in the specific heat capacity of atmospheric air. More precisely, increased atmospheric pollution results in decreased specific heat capacity, which, in turn, contributes to stronger winds. Consequently, the reduction in thermal capacity intensifies temperature fluctuations over the same time interval, meaning that temperature variations can shift between extreme values more frequently. Recently, a notable increase in wind intensity has been observed, contributing to destructive natural events. This aligns with the findings that lower thermal capacities are associated with intensified atmospheric instability and more frequent extreme weather conditions.

Acknowledgments. The study was funded by the A. Zlatarov University through project Ref. No. NIX-482/2023 "Application of Intuitionistic Fuzzy Logic in the Decision Making Process" and the project Ref. No NIX-486/2023 "Modeling Management Decisions with New Analysis Tools in a Fuzzy Environment".

References

1. Dhaka, S.K., Kumar, V.: Chapter 1 - Composition and thermal structure of the earth's atmosphere, In Earth Observation, Atmospheric Remote Sensing, Elsevier, pp. 1–18 (2023). ISBN 9780323992626, https://doi.org/10.1016/B978-0-323-99262-6.00023-7
2. Cimatoribus, A.A., van Haren, H.: Estimates of the temperature flux–temperature gradient relation above a sea floor. J. Fluid Mech. **793**, 504–523 (2016). https://doi.org/10.1017/jfm.2016.112
3. Laštovička, J., et al.: Global change in the upper atmosphere. Science **314**, 1253–1254 (2006). https://doi.org/10.1126/science.1135134
4. Danielson, E.W., Levin, J., Abrams, E.: Meteorology, p. 0073659630. McGraw Hill, ISBN (2003)
5. Lei, Y.: Extension of the reversible adiabatic equation to a general equation for ideal gas. Int. J. Thermophys. **42**(11), 157 (2021). https://doi.org/10.1007/s10765-021-02912-y
6. Bailyn, M.: A Survey of Thermodynamics. American Institute of Physics Press, New York p. 21 (1994). ISBN 0-88318-797-3
7. Tisza, L.: Generalized Thermodynamics. Cambridge. Massachusetts: MIT Press, p. 48 (1966)
8. Münster, A.: Classical Thermodynamics. Translated by Halberstadt, E. S. London: Wiley–Interscience. p. 45 (1970). ISBN 0-471-62430-6
9. Lindsey, R., LuAnn, D.: Climate Change: Global Temperature. Climate Change: Global Temperature | NOAA Climate.gov (2024)
10. Mihai, P.: Adiabatic thermodynamic manner of the description of the carbon dioxide pollutant excess of Earth's atmosphere. Int. Sci. J. Math. Model. (3), 106–111 (2021). WEB ISSN 2603–2929; PRINT ISSN 2535–0986, YEAR V
11. Last four years have been the warmest on record-and CO2 continues to rise, 07 Jan 2019, https://climate.copernicus.eu/, Accessed 1 Jan 2025
12. Terrie, K., Boguski, P.E.: Undesratnding units of measurements (2006). https://teesing.com/media/library/tools/understanding-units-of-measurement.pdf, Accessed 1 Jan 2025
13. Dıaz, B., Ramırez, J.E.: Adiabatic lapse rate of non-ideal gases: the role of molecular interactions and vibrations. Phys. Rev. E **102**, 042107 (2020)
14. Francesco Cairo, Atmospheric Thermodynamics (2010). https://doi.org/10.5772/19429 https://www.intechopen.com/chapters/21845, Accessed 1 Jan 2025
15. Atmospheric processes and phenomena, https://pressbooks-dev.oer.hawaii.edu/atmo/chapter/chapter-3-thermodynamics/, Accessed 1 Jan 2025

16. Gilchrist, J.D.: Fuels, furnaces and refractories, in international series on materials science and technology. Pergamon, pp. 347–353 (1977). ISSN 0362479X, ISBN 9780080204307
17. Spinoni, J., Formetta, G., Mentasghi, L., Forzieri, G., Feyen, L.: Global warming and windstorm impacts in the EU, EUR 29960 EN, Publication Office of the European Union, Luxembourg (2020). ISBN 978-92-76-12955-4, https://doi.org/10.2760/039014JRC118595

Enhancing Liquidity and Stability in DeFi: USD-collateralized Leveraged Restaking with Stablecoins

Petar Zhivkov(✉)

Inst. of I&C Tech., Bulgarian Academy of Sciences, Sofia, Bulgaria
petar.zhivkov@iict.bas.bg

Abstract. With the advancement of the Proof of Stake (PoS) Ethereum ecosystem, users can now restake their staked derivatives through protocols like EigenLayer and Symbiotic, receiving Liquid Restaked Tokens (LRTs) that represent their restaked assets and earning additional incentives. LRTs enhance the liquidity of restaked assets by enabling their use in secondary markets, such as for collateralized borrowing on Aave or for providing liquidity to yield markets like Pendle. This research proposes a novel approach to USD-collateralized leveraged restaking using stablecoins and explores the integration of Symbiotic's generalized restaking infrastructure, which expands the range of collateral to include any ERC-20 tokens. This approach offers a transformative shift in Actively Validated Systems (AVSs) by mitigating the economic volatility typically associated with cryptocurrency operations. The use of stablecoin derivatives ensures stable operational costs for AVS integrators and consistent yields for restakers, irrespective of ETH price fluctuations. This work provides a first step in the comprehensive analysis of liquid restaking.

Keywords: Proof of Stake · Liquid Restaking · Stablecoins

1 Introduction

The upgrade of the Ethereum blockchain to Proof-of-Stake (PoS) in 2022 represents a significant move towards a consensus mechanism that is more environmentally friendly and long-lasting [1,11]. The validator's stake is securely stored in escrow and may be reduced if any malicious conduct is identified, hence ensuring the system's crypto-economic security [3,8] and staking incentives are received when performing as expected. Nevertheless, engaging in solo staking requires a significant financial investment and a high level of technical proficiency in managing a validator node. In addition, staked ETH becomes less liquid throughout the staking period, which restricts its potential for other financial transactions. In order to address the significant barrier to entry, staking pools are implemented, enabling people to combine their ETH and jointly participate in staking. The management of these pools is entrusted to operators who oversee network validation and distribute incentives to members. Liquid Staked

D. Simian and L. F. Stoica (Eds.): MDIS 2024, CCIS 2486, pp. 200–215, 2025.
https://doi.org/10.1007/978-3-031-87386-7_14

Tokens (LSTs) are offered a direct and equal representation of staked ETH. The LSTs are interchangeable and may be traded. They represent the staked ETH and the rewards it generates. This enhanced the liquidity of the assets that were risked while also generating incentives. These coins have the potential to be used in several DeFi (Decentralized Finance) protocols, enabling stakers to increase their rewards and augment their revenues systematically.

Inspired by the success of staking, the restaking idea, developed by EigenLayer, involves a restake of previously staked assets, allowing stakers to receive more incentives and the protocol to enhance the security of decentralized applications. In this way, the concept of staking has been expanded to include not just the base-layer chains but also the smart contract layer, as shown by the practice of re-staking popularized by EigenLayer and later by Symbiotic. Both EigenLayer and Symbiotic are restaking protocols that serve as middleware on the Ethereum blockchain, facilitating enhanced security and additional benefits through staking and re-staking mechanisms. Staking involves users locking up their ETH to help secure the network, for which they receive compensation. Restaking extends this concept by allowing users to commit their already staked ETH to secure other initiatives beyond the Ethereum network, thus earning additional rewards. These initiatives are referred to as Actively Validated Services (AVSs), which are independent blockchain applications that leverage Ethereum's security infrastructure via the restaking protocol. In contrast to conventional staking methods, where staked ETH benefits solely the Ethereum blockchain, restaking or liquid restaking through restaking protocols extend security to both the Ethereum network and the associated AVS.

Liquid restaking tokens (LRTs) are intermediary protocols that are based upon the execution primitives and slashing offered by restaking protocols like EigenLayer and Symbiotic. Users deposit assets over the LRT protocol, which transfers those assets to operators, who then distribute the slashable collateral to a group of AVSs. The responsibilities of an LRT are finding prospective AVSs and negotiating security agreements with them in exchange for a certain share of supply and ongoing rewards—such as airdrops, inflation of AVS tokens, transaction fees, MEV, and protocol fees. After that, LRTs keep a service charge while returning the benefits to depositors. The internalization of risk within the LRT asset construction, depositor incentives, and AVS contract structure differ among these LRT systems. Despite the advancement of DeFi and restaking in particular, there are use of LRTs is yet limited and still there is a limited number of AVSs that participate in the process. However, with the use of new approaches to restaking the adoption of LRTs and involvement of AVSs is expected to expand.

Stablecoins are cryptocurrencies that are tied to a certain national currency, usually the US dollar. They are designed to facilitate more efficient transactions using non-stable cryptocurrencies compared to utilizing national currencies. They function on blockchains, decentralized digital ledgers where payments are authenticated and documented without requiring centralized settlement. Stablecoins achieve significantly reduced volatility compared to cryptocurrencies like Bitcoin (BTC) by maintaining a collateralized peg. The volatility of Bitcoin relative to the US dollar (USD) is around 10 times higher than that of major national currencies [2]. Stablecoin pegs, such as Tether, are solely maintained by private players, unlike fixed-rate national currencies. Additionally, unlike exchange-traded funds, there is no specific group of private participants designated in advance to fulfill this job.

In this research, we create a formal architecture for leverage stablecoin restaking using Restaked USDT (rsUSDT). rsUSDT secures, facilitates and monitors restaked network curation via Symbiotic. We undertake an analytical analysis to determine crucial indicators such as leverage staking multiplier and Health Factor (HF). To our knowledge, this is the first try to develop a leverage restaking strategy using stablecoin LRTs.

After providing a brief introduction, we will outline the article as follows: Sect. 2 will provide an in-depth literature review of liquid staking and restaking, stablecoins and find the gap in the literature. Section 3 introduces a more precise background of the problem, overviewing the EigenLayer and Symbiotic protocols. In Sect. 4, we introduce our model, followed by an analytical study in Sect. 5. Section 6 presents the discussion and outlines future research directions, while Sect. 7 concludes the paper.

2 Literature Review

A comprehensive literature review on staking and LSTs, DeFi lending, liquid restaking, and stablecoins is provided in this section.

Multiple experts have conducted research on the economic aspects of PoS staking. For instance, a study examined the impact of on-chain lending on the security of a Proof of Stake (PoS) blockchain [4]. Researchers discovered that when the return on loan contracts exceeds the inflation rate from staking, stakers are motivated to withdraw their staked tokens and lend them instead. Another recent research examined the Principal-Agent dilemma within the context of liquid staking and found the individuals deliberated on the quandary surrounding the decision between proportional representation and equitable punishment and put forward a specific strategy to demonstrate their lack of compatibility [12]. A study examined the difference in pricing of LSTs between the main and secondary market, known as the liquid staking basis or disparity [10]. It was noted that the liquid staking base expands when there is an increase in bitcoin volatility and a fall in liquidity in the secondary market.

Majority of leveraged staking positions (81,7%) outperformed traditional staking on Lido in terms of Annual Percentage Rate (APR) [14]. The study examines 442 leveraged staking positions over 963 days, involving a total volume of 537,123 ETH (approximately 877 million USD). Despite the promising returns, the study also identified substantial risks associated with leveraged staking, the Terra crash incident illustrated how token devaluation can drastically affect the market. The simulations suggested that leveraged staking could exacerbate the risk of cascading liquidations by introducing additional selling pressures from liquidation and deleveraging activities. Another study conducted a detailed investigation into the effects of the Ethereum merge on the DeFi lending platforms Compound and Aave [7]. Their study particularly focused on the strategies employed by Aave to mitigate the liquidation risks associated with collateralized stETH positions. Another recent work has developed a formal model for under-collateralized DeFi lending platforms, providing an empirical evaluation of the risks linked to leveraging, including impermanent loss, arbitrage, and liquidation [13].

Studies regarding liquid restacking are scarce due to their novelty in blockchain technology. As this innovative approach continues to evolve, more comprehensive research is expected to emerge, enriching our understanding and implementation of this mechanism. Yet some authors scratch the surface by investigating LSTs and LRTs withdrawal mechanisms [9]. According to the study, LRTs operate within a more intricate design framework, as they must manage withdrawals in alignment with various base protocols and services. The withdrawal processes for LRTs are complicated due to the diversity of rewards across protocols, which are issued in varying tokens and at different emission rates.

Literature Gap. At the time of this research, there are no studies available that specifically focus on the use of LRT tokens in money markets, particularly in relation to the usage of USD-pegged derivatives in restaking.

3 Background

This section aims to look further in-depth at the restaking ecosystem.

3.1 EigenLayer Protocol and the Use of AVSs

The introduction of PoS has significantly advanced blockchain technology, enhancing network security and scalability by allowing stakeholders to validate transactions proportionate to their holdings in Ether and their willingness to stake it. A notable development in this area is the emergence of restaking through EigenLayer. Since its introduction, EigenLayer has quickly climbed the Total Value Locked (TVL) rankings, achieving an impressive $20Bn in TVL by June 2024. This innovation enables ETH stakers to effectively leverage their assets twice over by validating not only Ethereum but also other compatible trust networks.

DeFi applications seeking to become a designated AVS must undergo a rigorous validation process. To date, 13 AVS are recognized within the EigenLayer ecosystem [6]. Once approved, these applications can capitalize on Ethereum's robust security framework, allowing developers to concentrate on enhancing their platforms.

With the implementation of the M2 upgrade, AVS operators now have the discretion to select specific AVS for validation. Similarly, restakers have the choice of which operators to entrust with their ETH.

It is essential to outline the main stakeholders within the AVS ecosystem:

- **AVSs:** These are decentralized applications (dApps) that utilize restaked ETH after being integrated into the EigenLayer system.
- **Validators:** These nodes are central to the functioning of EigenLayer's restaking mechanism, supporting the security and operation of AVS.
- **AVS Operators:** These are individuals or entities that manage nodes capable of being penalized (slashable), in return for predetermined rewards. AVS Operators manage Validators to bolster the security and functionality of AVS on EigenLayer.

AVSs and restaking are still a relatively new and immature concept within the DeFi space. The revenue models for AVSs are not fully established, making it difficult to predict future returns accurately. The revenues from AVSs depend on on several unpredictable variables such as user adoption rates, transaction volumes, and evolving regulatory landscapes that can influence the profitability of AVSs. Despite its solid TVL, the design of EigenLayer's protocol and its tokenomics have highlighted several challenges and controversies.

3.2 Symbiotic Protocol and the Use of ERC-20 for Restaking

Expanding on the idea of securing AVSs through liquid restaking, the recently introduced platform Symbiotic allows for the use of any ERC-20 asset as collateral for restaking and this means they are directly compatible with LSTs like Lido's stETH. While EigenLayer receives only ETH and this makes the main difference between the two protocols. Developers may use a diverse range of assets to safeguard their apps due to the extensive customization choices and adaptable restaking model available. Symbiotic's introduction into the realm of shared security, with a "stake anything" concept, is likely to disrupt the current landscape.

The agnosticism of Symbiotic is very important destinction as it allows EigenLayer LRTs to also participate in Symbiotic. This results in double slashing, thereby creating a double yield tranche. In addition, it allows dollar-pegged derivatives to enter the restaking ecosystem.

3.3 Types of Stablecoins

Stablecoins are unique types of tokens that are anchored to a base currency, known as a numeraire [5]. Typically, stablecoins are linked to stable assets like the U.S. Dollar. We are outlying the 3 main factors defining stablecoins.

Factor 1 Trust. Trust in stablecoins exists in two dimensions: centralized and decentralized. A stablecoin is either issued by a centralized entity that requires trust or governed by smart contracts that operate in a decentralized and trustless manner.

Factor 2 Mint or Redeem Mechanism. Stablecoins can be categorized into three distinct mint or redeem mechanisms:

- **Straight Up: Exchange for $1** This is the simplest mechanism where one stablecoin can always be minted by providing $1 and redeemed for $1. Arbitrage ensures the stablecoin price remains at $1. If the stablecoin trades above $1, people mint and sell it, increasing the supply until the price drops to $1. Conversely, if it trades below $1, people buy and redeem it, decreasing the supply until the price rises to $1.
- **Assets: Exchange for $1 Worth of Assets** This mechanism is similar to "Straight Up," but instead of providing $1, you provide $1 worth of eligible assets to mint one stablecoin. Likewise, one stablecoin can be redeemed for $1 worth of assets. Arbitrage operates similarly: if the stablecoin trades above $1, people mint and sell it; if it trades below $1, they buy and redeem it. The key difference is that the stability of the eligible assets is crucial to prevent a 'death spiral.'
- **Over-Collateralization: Lock Up More Than $1 in Assets to Mint the Stablecoin** The over-collateralized mint and redeem mechanism is more complex than the previous methods. To issue one stablecoin unit, more than $1 worth of eligible collateral assets must be secured within a protocol. The collateralization rate, which is the value of the collateral divided by the number of stablecoin units issued, must exceed 1. If this rate drops below a certain threshold, an auction is triggered. Participants bid stablecoins for the collateral, and the winning bid's stablecoins are burned to maintain over-collateralization. The collateral can always be redeemed for the initial issuance amount of the stablecoin (excluding any potential interest payments). If the stablecoin is over-collateralized but trades below $1, users are likely to buy and redeem it for collateral. This action contracts the stablecoin supply until its price returns to $1. As long as over-collateralization is maintained, a low price is unlikely to persist. Thus, when the value dips below $1, it's a good time to repay debt, contracting the supply until parity is restored. However, over-collateralization alone isn't sufficient to maintain a strict $1 peg. For instance, the decentralized stablecoin DAI uses this mechanism but requires additional measures to stay stable at $1.

Factor 3 Backing. Backing identifies the assets that support a stablecoin's value. We categorize stablecoin backing into four groups:

- **Non-crypto assets** - These are real-world assets with stable values, like bank deposits, cash, and treasury bills.
- **Crypto assets** - This includes various types of cryptocurrencies.
- **Flexible supply crypto assets** - These cryptocurrencies have a supply that decreases when new stablecoins are created and increases when stablecoins are redeemed.

- **Carry trade** - This involves selling futures contracts while purchasing the underlying asset. If the futures are priced higher, this strategy can generate a "guaranteed" profit.

There are four distinct combinations of elements that have historically been utilized to create stablecoins. These combinations are termed as centralized asset-backed, decentralized crypto-over-collateralized, decentralized algorithmic, and centralized carry trade-backed. Table 1 below outlines the trust, mint or redeem, and backing mechanisms for these four stablecoin types.

Table 1. Types of stablecoins.

Type	Centralization	Mint/Redeem Mech	Backing
Centralized asset-backed	Centralized	Straight up: Exchange for $1	Non-crypto assets
Decentralized crypto-over-collateralized	Decentralized	Over-collateralization	Crypto assets
Decentralized algorithmic	Decentralized	Assets: Exchange for $1 worth of assets	Flexible supply crypto assets
Centralized carry trade-backed	Centralized	Assets: Exchange for $1 worth of assets	Carry trade

According to DefiLlama, by the time of writing, centralized asset-backed stablecoins account for 93% of the market capitalization. Decentralized crypto-over-collateralized stablecoins account for 4%, while centralized carry trade-backed stablecoins make up 2.5%. Decentralized algorithmic are counting for just 0.5% of the stablecoin market capitalization.

3.4 Stablecoins for Restaking

LRTs are often tied to the native currency of the PoS blockchain such as ETH or SOL. Issuers of these pegged tokens guarantee redemption of each token for an equivalent amount of the numeraire asset, a promise they generally support by holding reserves that are equal to or greater than the total issued tokens.

In the paper we propose to use dollar-pegged ERC-20 s for restaking derivatives through the Symbiotic infrastructure, which aims to offer several key benefits:

- **Stable Economic Security:** With dollar-pegged collateral, the economic security provided to the networks being restaked remains stable regardless of fluctuations in the price of ETH. This ensures a consistent level of security that is not impacted by volatile crypto markets.

- **Fixed Operational Expenses:** For the Actively Validated Services (AVSes) being secured by the restaked collateral, the operational costs are fixed in USD terms. This provides predictability and allows them to budget more effectively.
- **Fixed Yields for Restakers:** Similarly, the yields earned by restakers are also pegged USD-derivative, giving them a stable return on their capital regardless of ETH price fluctuations.

In contrast, using ETH or other crypto-native collateral can expose the restaking ecosystem to greater volatility and unpredictability. This is illustrated in Fig. 1, which shows the price of ether.fi's LRT - weETH. The token weETH had a relatively stable price correlation with ETH over the past three months, with the pair weETH/ETH trading between 1.027 and 1.048, a difference of just 2%. However, when compared to USD, the weETH/USD price fluctuated from \$2975 to \$4100, representing a 27.4% difference. This market volatility is beyond the control of the protocols. Symbiotic's support for dollar-pegged stablecoins helps mitigate these risks, creating a more stable, reliable, and efficient restaking infrastructure.

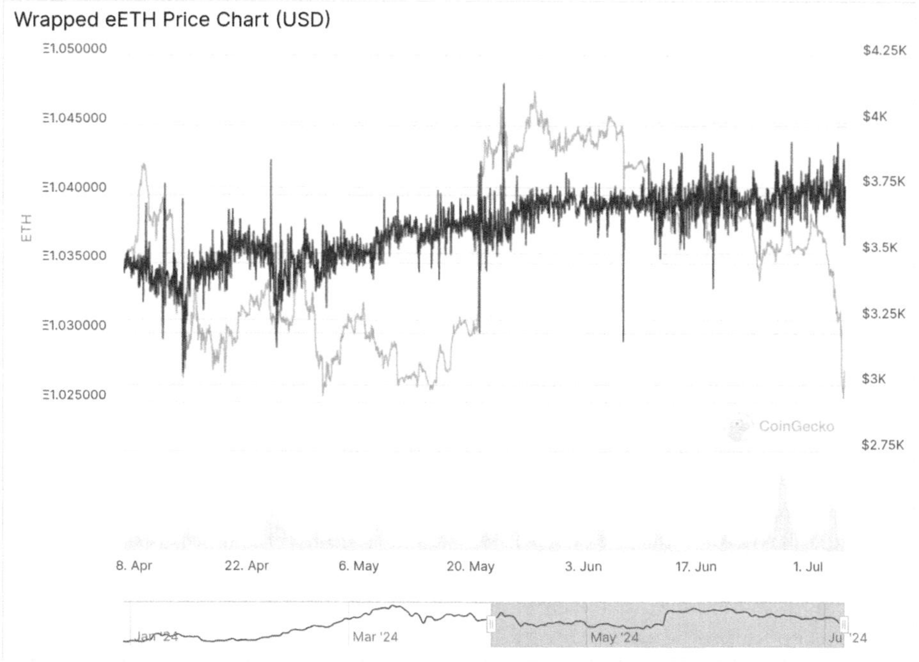

Fig. 1. The figure shows 3-month price ratio for the ether.fi's LRT weETH. The purple line shows the pair weETH/ETH. The green line shows the weETH/USD price. (Color figure online)

4 Model

The current model is influenced by the approach outlined in previous work [14] that overviews leveraged staking with ETH-based derivatives. This paper introduces several key differences. First, rather than focusing on staking, we explore restaking and the leverage of LRTs. Second, the model employs a USD-pegged derivative instead of an ETH-pegged derivative, resulting in lower volatility, which positively impacts the Loan-to-Value (LTV), liquidation rates, and overall HF. Finally, we introduce the usability of LRTs in the yield market, specifically through integration with Pendle to enhance yield generation.

4.1 System Participants

This subsection outlines the various participants involved in the system and their respective roles.

- **Users:** In this system, User U_i is represented as capable of connecting to various DeFi systems. U_i may use a variety of tactics and options to optimize financial rewards.
- **Restaking Protocols:** U_i may restake ERC-20 tokens on liquid restaking platforms (e.g., Symbiotic) to earn LRTs (e.g., reUSDT). Later these LRTs can be used for further rewards by using them for liquidity provision, collateralized borrowing, or simply swapping them for other tokens.
- **DeFi Providers:** U_i may provide a single asset to borrowing and lending platforms such as Morpho, Curve, or Aave and use it as collateral to get a loan in the form of another asset. They can also use them to provide liquidity for more advanced DeFi protocols such as Pendle.

4.2 Levareged Borrowing

This process involves users initially exchanging asset A for B within a DEX pool. Next, asset B is used as collateral on lending platforms to borrow asset A. The borrowed asset A is then exchanged again for B in the DEX pool, allowing users to iteratively increase their leverage borrowing positions. Most lending pools involve different assets for asset A and asset B. However, in 'single asset leverage borrowing,' assets A and B are the same. This eliminates the need for DEX exchanges, allowing the leverage loop to continue without additional swapping. The financial incentive behind this strategy lies in the user's ability to expand their deposit-borrow leverage by repeatedly swapping, depositing, and borrowing. Even though the deposit rate offered by the lending platform is often lower than the borrowing rate, the larger total amount of asset B deposited compared to asset A borrowed generally results in net earnings.

LTV is a financial term that represents the ratio of a loan to the value of the collateral backing the loan. It is commonly used in both traditional finance and DeFi. In the context of DeFi, LTV indicates the percentage of an asset's value that a borrower can borrow against. For example, an LTV of 70% means that for every $100 worth of collateral (such as ETH or stablecoins) deposited, the borrower can take out a loan of up to $70. A higher LTV ratio means the borrower can borrow more relative to the collateral value, but it also increases the risk of liquidation if the collateral value falls. The LTV ratio depends on the platform and the specific risk models they use, but usually for ETH and the LTV ratio is between 65–75% as a more volatile asset and for stablecoins is between 75–85% as the liquidation threshold for stablecoins is also usually higher because the volatility of the assets is less compared to ETH.

4.3 Levareged Restaking

U_i first restakes a principal amount of R of USDT on Symbiotic at time t_0 to acquire $\frac{R}{p_{t_0}^{rs}} = rsUSDT$, where $p_{t_0}^{rs}$ denotes the rsUSDT to USDT price and is equal to 1 in the primary market. Next, U_i supplies rsUSDT on Aave to borrow $R \cdot l \cdot p_{t_0}^{a}$ amount of USDT, where l denotes the LTV ratio and $p_{t_0}^{a}$ denotes the rsUSDT to USDT price used by the Aave V2 lending protocol (Fig. 2).

Then U_i restakes the borrowed USDT on Symbiotic. U_i performs this loop multiple times, amplifying both the restaking reward from Symbiotic and the deposit-borrow revenue from Aave 4.

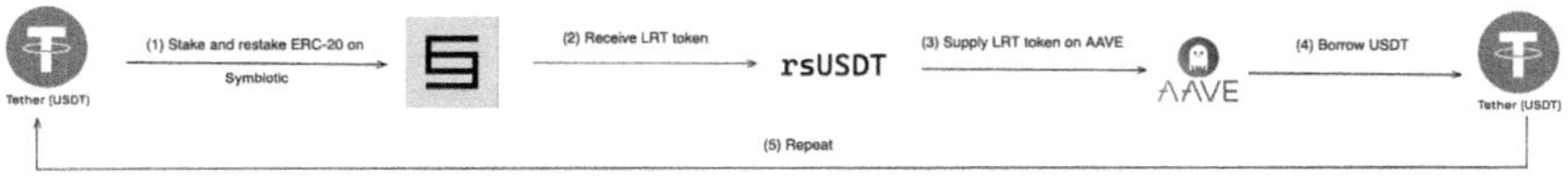

Fig. 2. An illustration of direct leverage restaking strategy.

4.4 Understanding Leverage Borrowing and Leverage Restaking

While leverage restaking and leverage borrowing both follow iterative patterns, they differ in several. Firstly, leverage restaking primarily targets LRTs, whereas leverage borrowing encompasses a broader range of tokens. Secondly, leverage restaking aims to enhance both staking and restaking rewards, along with deposit-borrow revenue, whereas leverage borrowing concentrates solely on boosting deposit-borrow revenue. Thirdly, they entail different risk factors. In addition to market risk, leverage restaking involves the potential for slashing penalties. Therefore, it is important AVSs to go through a thorough validation process.

In summary, although the traditional leverage borrowing strategy bears some similarities to the leverage restaking strategy discussed here, they diverge in terms of the types of assets involved, revenue sources, and associated risks. This approach not only offers increased rewards for stakers but also enhances the overall security and resilience of the DeFi ecosystem by expanding the participation of assets in the restaking process. Consequently, restaking emerges as an attractive option for users seeking to maximize their returns while bolstering the stability of decentralized applications.

5 Analytical Study

This section conducts an analytical study on the leveraged liquid restaking strategy using the Symbiotic restaking protocol. We assume that U_i can complete n loops within a short time interval such that the rsUSDT price remains unchanged as shown in Fig. 3.

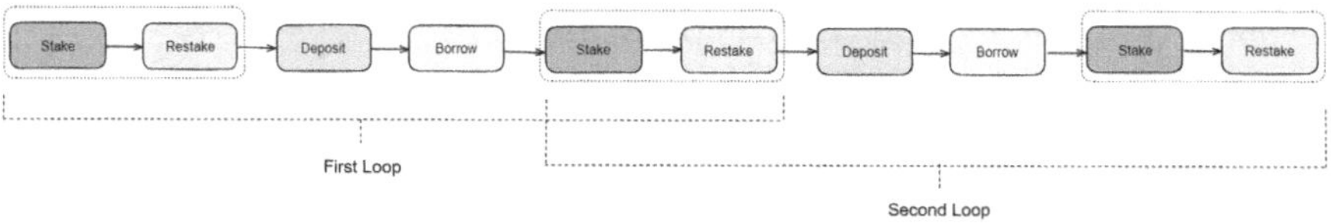

Fig. 3. Illustration of direct leverage restaking loops. The user U_i makes n loops following the line of acts $stake, restake, deposit, borrow, stake, restake$. Each loop should finish with stake, restake action.

As a rational participant, U_i determines the value of n according to its risk profile. Let $p^a_{t_0}$ denote the Aave lending price of rsUSDT and $p^m_{t_0}$ be the rsUSDT to USDT market price. U_i acquires a total investment amount of $A(R, n)$ USDT, collateralized by the restaked tokens on the Symbiotic protocol, and completes the loop n times.

$$A(R, n) = R \cdot \sum_{k=1}^{n+1} \left(\prod_{i=1}^{k} r_i \right) \tag{1}$$

Here, r_i represents the percentage of rsUSDT restaked in each loop. This process amplifies both the restaking rewards from the Symbiotic protocol and the deposit-borrow revenue from Aave. As we talk about stablecoins and USDT in particular, this should be in the range of

5.1 Leverage Multiplier

Assume U_i invests a total asset (restaked USDT on Symbiotic) of $A(R, n)$ USDT through leveraged liquid restaking with an initial investment of R USD, the leverage multiplier $\mathrm{LevM}(R, n)$ is defined as:

$$\mathrm{LevM}(R, n) = \frac{A(R, n)}{R} \tag{2}$$

5.2 Direct vs Indirect Leveraged Restaking

Both direct and indirect leveraged restaking strategies are designed to amplify restaking rewards through a recursive methodology. However, direct leveraged restaking acquires rsUSDT from the primary market, whereas indirect leveraged restaking obtains rsUSDT from the secondary market. Utilizing the same principal amount of USDT, U_i generally acquires more rsUSDT through the indirect strategy as the rsUSDT to USDT price in the secondary market is often below 1. However, this approach incurs additional costs associated with swaps and is susceptible to potential price slippage.

5.3 Health Factor and Leveraged Restaking Formula

The Health Factor is a crucial metric used to monitor the status of each position in leveraged restaking. A position is susceptible to liquidation if its HF drops below 1. HF is calculated based on the value of the collateral and the debt, indicating the safety of the user's position. Parameters such as the loan-to-value ratio (l) and the liquidation threshold (LT) are essential for determining the HF. For example, a position must maintain a certain price threshold to remain secure (HF $>$ 1), which is dependent on the percentage change in the price of the restaked asset.

For leveraged restaking using Symbiotic, the leverage multiplier (LevM) is defined as the ratio of the total asset amount ($A(R, n)$) to the initial restaked amount (R). The following equations illustrate the leverage multiplier for both direct and indirect leveraged restaking scenarios:

Direct Leveraged Restaking. In direct leveraged restaking, the primary market price $p_{\mathrm{m,t0}} = 1$. The leverage multiplier (LevM) is given by:

$$\mathrm{LevM}(R, n) = \frac{1 - (l \cdot p^a_{t_0})^{n+1}}{1 - l \cdot p^a_{t_0}} \tag{3}$$

where:

- R is the initial restaked amount.
- n is the number of restaking loops.
- l is the loan-to-value ratio.
- $p^a_{t_0}$ is the rsUSDT to USDT lending price

Indirect Leveraged Restaking. In indirect leveraged restaking, when $l \cdot p^a_{t_0} < p^m_{t_0}$, the leverage multiplier (LevM) is calculated as:

$$\mathrm{LevM}(R, n) = \frac{1 - (l \cdot p^a_{t_0}/p^m_{t_0})^{n+1}}{1 - l \cdot p^a_{t_0}/p^m_{t_0}} \tag{4}$$

where:

- R is the initial restaked amount.
- n is the number of restaking loops.
- l is the loan-to-value ratio.
- $p^a_{t_0}$ is the rsUSDT to USDT lending price
- $p^m_{t_0}$ is the rsUSDT to USDT market price.

Equation (3) is a special case of Eq. (4), because in direct leveraged restaking $p^m_{t_0} = 1$.

5.4 Pendle V2 Integration for Stablecoin LRTs

The integration of the Pendle V2 model, which allows for the tokenization of future yield, provides a solution for transforming rsUSDT into Principal Tokens (PTs) and Yield Tokens (YTs), thus enabling the creation of a structured money market. This section elucidates the mechanism of applying Pendle V2 to rsUSDT, presenting a formalized approach to trading and earning additional yield on these tokens, and demonstrates the economic advantages through mathematical formulations and theoretical models.

Pendle V2's core innovation lies in the tokenization of yield-bearing assets. For rsUSDT, this involves converting each rsUSDT into a pair of tokens: Principal Tokens (PTs), which represent the principal value of the rsUSDT and are redeemable at maturity, and Yield Tokens (YTs), which represent the yield generated by the rsUSDT until maturity, see Fig. 4.

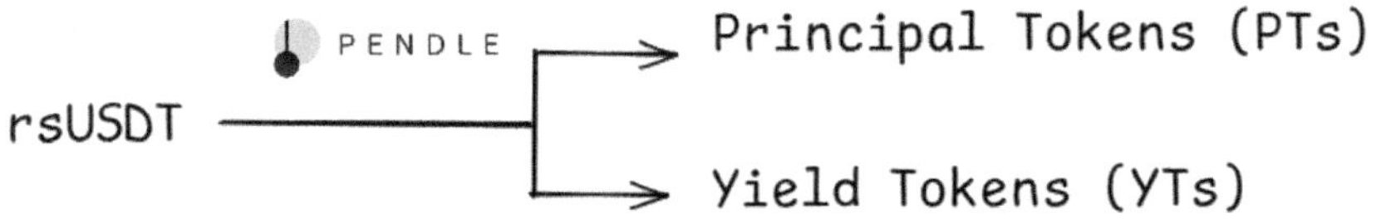

Fig. 4. The separation of rsUSDT into two components: Principal Tokens (PTs), which hold the principal value and can be redeemed at maturity, and Yield Tokens (YTs), which capture the yield produced by rsUSDT until maturity.

Pendle V2 employs an Automated Market Maker (AMM) model to facilitate the trading of PTs and YTs. This model determines the price of PTs based on the formula:

$$price_{PT}(t) = \frac{\ln\left(\frac{p(t)}{1-p(t)}\right)}{rateScalar(t)} + rateAnchor(t) \tag{5}$$

where $price_{PT}(t)$ is the price of the principal token at time t, $p(t)$ is the pool utilization ratio at time t, $rateScalar(t)$ adjusts for capital efficiency, and $rateAnchor(t)$ is the base interest rate, dynamically adjusted to reflect market conditions. This model allows users to engage in the trading of PTs and YTs, thereby establishing a liquid market for rsUSDT.

5.5 Application Example

To illustrate the application of Pendle V2, consider a user U_i who converts their rsUSDT into PTs and YTs. The process is as follows:

1. Tokenization: U_i converts $rsUSDT$ into PT and YT using the Pendle V2 protocol:

$$rsUSDT, Pendle\{PT_{rsUSDT}, YT_{rsUSDT}\} \tag{6}$$

2. Trading and Yield Earning: U_i trades the PTs (PT_{rsUSDT}) and YTs (YT_{rsUSDT}) on the Pendle V2 AMM. The price of PTs at time t is given by:

$$price_{PT}(t) = \frac{\ln\left(\frac{p(t)}{1-p(t)}\right)}{rateScalar(t)} + rateAnchor(t) \tag{7}$$

3. Economic Advantage: By holding YTs, U_i earns additional yield until the maturity of the PTs. The total yield earned can be expressed as:

$$Yield_{total} = R \cdot \mathrm{YT}_{yield} \tag{8}$$

where YT_{yield} is the yield generated by the YTs over the period.

This structured approach provides U_i with a flexible and efficient way to manage and maximize the returns on their rsUSDT.

6 Discussion and Future Research

Our study on USD-pegged leveraged restaking with stablecoins presents several implications for the DeFi ecosystem. The integration of Symbiotic's generalized restaking infrastructure, which supports any ERC-20 token as collateral, provides a significant advancement over existing protocols that predominantly rely on ETH. This innovation could lead to a more robust and diversified DeFi landscape by enhancing liquidity and stability through the use of fiat-pegged stablecoins. One of the key findings is the mitigation of economic volatility typically associated with cryptocurrency operations. By using stablecoin derivatives, AVS integrators can ensure stable operational costs and consistent yields for restakers, irrespective of ETH price fluctuations. This stability is crucial for attracting more participants into the DeFi space, particularly those who are risk-averse.

As potential obstacles for the model, there is a risk of funds being lost either through liquidation, as a result of leveraging, or through slashing during restaking. Therefore, it is important to carefully select AVSs and actively manage the HF to mitigate these risks.

Future research should focus on several areas to build upon our findings. First, a deeper empirical analysis of the long-term performance of USD-pegged leveraged restaking compared to traditional methods is necessary. This would involve assessing the stability of yields, the resilience of the system under market stress, and the overall impact on the DeFi ecosystem. Additionally, exploring the regulatory implications and recognizing the associated risks. Conduct stress tests to simulate various extreme scenarios. These tests reveal that leveraged liquid restaking significantly increases the risk of cascading liquidations within the Symbiotic ecosystem because it triggers additional selling pressures during liquidations and deleveraging.

7 Conclusion

This paper has introduced a novel approach to USD-pegged leveraged restaking using stablecoins within the Ethereum ecosystem. By leveraging the Symbiotic protocol's ability to utilize any ERC-20 token as collateral, we have demonstrated a method that enhances the liquidity and stability of restaked stablecoins. The proposed approach mitigates economic volatility, ensuring stable operational costs for AVSs. While yield variability may depend on the specific restaking protocol and its tokenomics, the approach aims to promote more predictable and stable compensation for restakers during broader market fluctuations. In addition, the use of USD-pegged derivatives is expected to result in a lower volatility, which impacts the LTV, liquidation rates, and overall HF, compared to ETH-pegged ones.

Our findings highlight the potential for this method to transform AVSs by providing a more stable and predictable financial environment. The integration of stablecoin derivatives offers significant benefits over traditional crypto-native collateral, particularly in reducing the risks associated with market volatility. As the DeFi landscape continues to evolve, it is imperative that future research explores the long-term performance and regulatory implications of USD-pegged leveraged restaking. By addressing these areas, we can further refine and enhance the stability and inclusivity of decentralized financial systems. In summary, the adoption of stablecoin derivatives in restaking protocols represents a significant advancement in DeFi, offering a promising pathway to more resilient and accessible financial ecosystems.

Acknowledgement. The work is supported by the National Scientific Fund of Bulgaria under the grant DFNI KP-06-N52/5.

References

1. Agrawal, S., Neu, J., Tas, E.N., Zindros, D.: Proofs of proof-of-stake with sublinear complexity (2022). arXiv preprint arXiv:2209.08673
2. Baur, D.G., Dimpfl, T.: The volatility of bitcoin and its role as a medium of exchange and a store of value. Empirical Econ. **61**(5), 2663–2683 (2021)
3. Buterin, V., et al.: Combining ghost and casper (2020). arXiv preprint arXiv:2003.03052
4. Chitra, T.: Competitive equilibria between staking and on-chain lending. Cryptoeconomic Syst. **1**(1) (2021). https://doi.org/10.21428/58320208.9ce1cd26
5. Cintra, T.N., Holloway, M.P.: Detecting depegs: towards safer passive liquidity provision on curve finance (2023). arXiv preprint arXiv:2306.10612
6. Eigenlayer Team: Eigenlayer AVS. https://www.eigenlayer.xyz/ecosystem?category=AVS. Accessed 15 July 2024
7. Heimbach, D., et al.: The impact of the ethereum merge on market liquidity. In: Financial Cryptography and Data Security (2022)
8. Neu, J., Tas, E.N., Tse, D.: Ebb-and-flow protocols: a resolution of the availability-finality dilemma. In: 2021 IEEE Symposium on Security and Privacy (SP), pp. 446–465. IEEE (2021)

9. Neuder, M., Pai, M., Resnick, M.: Optimizing exit queues for proof-of-stake blockchains: a mechanism design approach (2024). arXiv preprint arXiv:2406.05124
10. Scharnowski, S., Jahanshahloo, H.: The economics of liquid staking derivatives: basis determinants and price discovery. Available at SSRN 4180341 (2023)
11. Tang, W., Yao, D.D.: Transaction fee mechanism for proof-of-stake protocol (2023). arXiv preprint arXiv:2308.13881
12. Tzinas, A., Zindros, D.: The principal–agent problem in liquid staking. In: International Conference on Financial Cryptography and Data Security, pp. 456–469. Springer (2023)
13. Wang, Z., Qin, K., Minh, D.V., Gervais, A.: Speculative multipliers on DeFi: quantifying on-chain leverage risks. In: Financial Cryptography and Data Security (2022)
14. Xiong, X., Wang, Z., Chen, X., Knottenbelt, W., Huth, M.: Leverage staking with liquid staking derivatives (lsds): opportunities and risks (2023). arXiv preprint arXiv:2401.08610

Security and Cryptography in Dynamic Systems

Comprehending Steganography Through A Comparative Analysis with Watermarking: A Study

Cristiana Constantinescu(✉)

Faculty of Automatics, Computer Science and Electronics, University of Craiova, Doctoral School 'Constantin Belea', Craiova, Romania
constantinescu.cmaria@gmail.com

Abstract. Steganography and Watermarking represent two key pillars of data hiding and privacy protection. Despite the particular advancement of digital approaches and the expansion of tailored applicability scenarios for each of the two that have occurred in the last decades, steganography and watermarking are often times mistaken to be one and the same science. Within this paper, we acknowledge the importance and magnitude of each and aim to trace clear borders between the two, in accordance to nowadays state of art in data hiding. We comply and accept that in the recent past steganography and watermarking used to be more tangential and in conjunction. However, we take into accountancy that nowadays the two have individually evolved in separate manners. Within this paper, we propose and analyze what makes them apart: the discrepancy elements between the two, with a more thorough perspective on Steganography.

Keywords: Steganography · Data protection · Information Security · Watermarking · Data hiding · Image Steganography · Steganalysis · CIA Triad · Digital Content Security

1 Introduction

Privacy and confidentiality are two of the important virtues humans oriented to and addressed in the context of present state of computer technology. European regulations (EU GDPR) adopted in this regard, trails addressing users' data and privacy, leaks and unauthorized data exchange (Cambridge Analytica, the lawsuit against Meta, Google, TikTok, Snap in regard to children safety etc.) stand as a testify for the importance and impact of this subject.

Computer science has emerged and is now available to almost any individual. For some of us, day-to-day workplace and education have moved to interconnected computers. Social contexts such as pandemics, conflicts have once again reinforced the necessity and importance of having computer solutions capable to handle and protect users' data. These situations imposed a greater interest in

D. Simian and L. F. Stoica (Eds.): MDIS 2024, CCIS 2486, pp. 219–238, 2025.
https://doi.org/10.1007/978-3-031-87386-7_15

data protection. One of the mostly used protection technique is data encryption, therefore cryptography. These unfortunate contexts led our society to actually live the theoretical approach of cryptography principles, where sensitive data is exchanged across insecure channels. Therefore, as currently used techniques fail to properly handle unauthorized data leaks and access, there is obvious that new approaches must be studied in order to provide more competent solutions in this specific contextual requirements (where data are never secured). One of the applicable solution lays in steganography. However, steganography is often times confused with watermarking, making it not seem an obvious feasible solution at the first glance. Further in this paper we will show how steganography is, in fact, applicable and feasible for this matter.

This paper offers an in depth inquiry on digital steganography, identifies dilemmas, approaches, assessments and methodology of the subject, as well as it contrasts steganography with cryptography and watermarking. Based on this study, we have identified new criteria that we advise to be seriously taken into consideration when addressing steganography in modern context. It is a must for a proper steganography solution in this matter to assure secure communication between peers using any *insecure, unmanaged (by the recipients)* communication channel, from institutional emails to direct messaging clients. In order to properly asses a relevant role to steganography in the field of modern computer communication, one must correctly comprehend end evaluate computer security principles and methodology that this paper further addresses.

2 Placing Steganography and Watermarking on Information Security's Dashboard

As steganography is often mistaken to watermarking, the CIA triad comes in help in order to asses the relevance of steganography and to acknowledge its proper place in computer security field. Information Security, IT Security and Cybersecurity are other definitions that are often misinterpreted as being the same. In fact, they are all different. Therefore, each of the three represents an individual entity between which the following inclusion relationship takes place: $\text{CS} \subset \text{ITS} \subset \text{IS}$, where CS stands for Cybersecurity, ITS for IT Security and IS for Information Security. Following up, we will refer to the least three as *security fields* or *fields*. The CIA triad represents the **C**onfidentiality, **I**ntegrity and **A**vailability aspects applied to each field in correspondence to each of the field's characteristics. Having this as a starting point, it is now important to place cryptography, watermarking and steganography on the dashboard of security.

The following diagram illustrates how IS, ITS, CS, cryptography, steganography and watermarking are related.The terms under which cryptography and steganography can overlap each other is following discussed within this paper. **Cryptography** is a science that represents the practice and study of techniques used to assure the protection of data in terms of integrity, availability, non-repudiation and confidentiality. **Watermarking** represents the practice and study of embedding data inside or perceptual *on top* of other data, with the intent to clearly and recognizably mark origin, appurtenance, ownership or other such information about a data representation (paper document, photography, sound etc.). As watermarking should alter the data it is attached on in a small measure, it primarily contributes to the integrity of data. Watermarking can both be applied to data at rest and in transit (Fig. 1).

Fig. 1. Dashboard of Security Diagram.

Steganography is a science that represents the practice and study of embedding data inside other data, in an imperceptible manner, in order to assure the principles of availability, integrity, and confidentiality of the embedded data. At some extent, steganography can even nowadays be applied to Information Security. It is generally accepted to be applied to Cybersecurity also. Given the concept of embedding data inside other data, *watermarking is part of steganography.*

3 Generalized Concepts (GC). Types Categorization (TC). Goals and Objectives (GO)

GC of Steganography Steganography has been present as valuable method of information protection even from the ancient times, as Survey [9] discussed.

A *steganographic system* is represented by the 5-tuple of ($\mathcal{D}$, $\mathcal{D}'$, $\mathcal{P}$, $\mathcal{H}$, $\mathcal{U}$), where $\mathcal{D}$ represents a finite set of the plain data, in clear - set that contains both the secret that needs to be protected and the mask that has to withhold the secret, expressed in its readable, comprehensive form. When the secret has a different representation from the mask, (e.g.: secret is sound, mask is image), the $\mathcal{D}$ set can be further divided in correspondent data subsets; $\mathcal{D}'$ represents a finite set of possible obtained representations of the $\mathcal{D}$ set - in a comprehensive form of its own, but different from the original data, $\mathcal{P}$ represents a pattern in which the clear data is hidden inside the representation of the steganographed data, decided by the steganographer, $\mathcal{H}$ represents a function used for hiding, dependent on each p$\in\mathcal{P}$, that morphs the clear data into its corespondent, $\mathcal{U}$ represents a function used for hiding, dependent on the same each p$\in\mathcal{P}$, that retrieves the data in its original shape.

A *Steganographic Mechanism* is each 3-tuple ($\mathbf{p}$, $\mathbf{h}$, $\mathbf{u}$), with $\mathbf{p} \in \mathcal{P}$, $\mathbf{h} \in \mathcal{H}$, $\mathbf{u} \in \mathcal{U}$ of a steganographic system, alongside any other potential privacy, security and data manipulation techniques additions the steganographer might decide upon. *Secret Message* $\in \mathcal{D}$, *Cover Media* $\in \mathcal{D}$, *Stego-Object* $\in \mathcal{D}'$.

The *Stego-Object* must have a meaning of its own and must not betray the presence of foreign data within. The presence can be signaled visually, by having a *Stego-Object* with visible signs of data alteration, such as but not limited to added noise (for both sound and image steganography - *see* Sect. 3), distorted pixels (for image steganography- *see* Sect. 3), distorted audio-gram or any potential data corruption that could be perceived by the human sensors [9–11].

Upon extraction (Fig. 2a), the *Cover Media* can and should be ignored, meaning it should not be reconstructed due to security measures, raising difficulties in an attacker's progress by rendering the original resources unavailable.

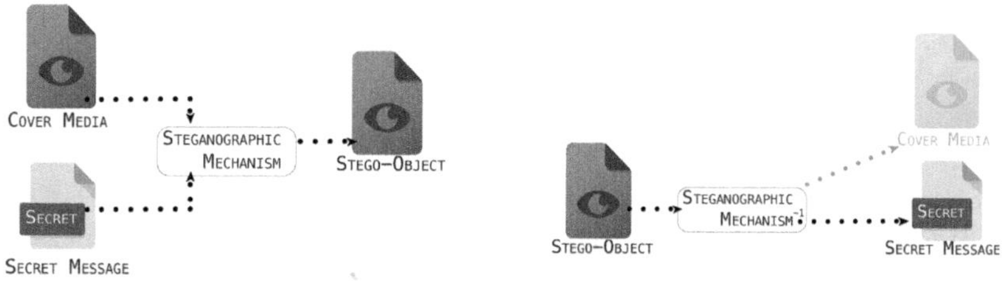

a) Encoding the Secret. b) Decoding the Secret.

Fig. 2. Generalized Concept of Steganography.

GC of Watermarking. While steganography is focused on confidentiality, watermarking's focus is inclined to recognition and property marking, with the intention of preserving data integrity and ownership, as discussed in [Sect. 2].

Watermarking implies a *stamping-like* action upon the data that it is applied to. The watermark should posses certain characteristics, among which: **uniqueness**- each watermark should be visibly distinct from any other watermark, **recognisability**- based on their uniqueness, watermarks facilitate the identification of the entity they were issued for, making a clear statement regarding the ownership or any other appurtenance that the content they are applied to has, **robustness** implies the resilience of the watermarks against attacks (intentional or not) that target the tampering or removal.

A simple *watermarking system* is represented by the reduced 3-tuple of ($\mathbf{D}$, $\mathcal{W}$, $\mathcal{E}$), where $\mathbf{D}$ represents a finite set of original data, $\mathcal{W}$ represents a finite set of possible representations of the watermark, $\mathcal{E}$ represents a set of functions used for embedding w$\in\mathcal{W}$, upon each d$\in\mathcal{D}$. A *Watermarking Procedure* is represented by the embedding function $e \in \mathcal{E}$ and oftentimes implies the overlaying of the *Watermark Data* on top of or within the representation of the *Original Data.* The *Watermarked Data* represents the function's $e \in \mathcal{E}$ output. For particular cases in which the watermark needs to be extracted and further analyzed, the 3-tuple can be extended with a function dedicated to the extraction and the function $e \in \mathcal{E}$ should be adjusted in order to preserve the *Watermark Data*'s integrity upon embedding (Fig. 3).

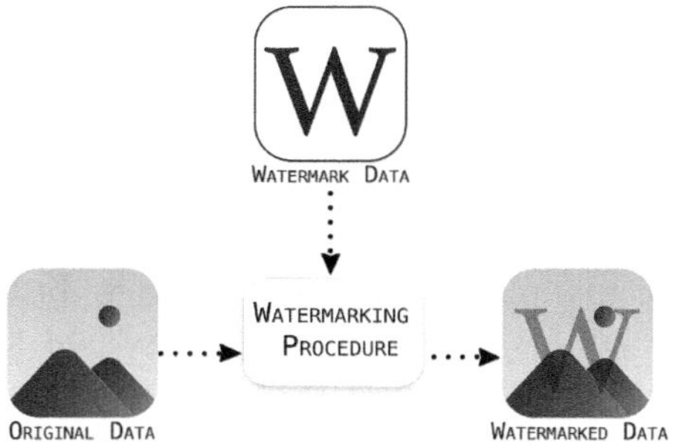

Fig. 3. Generalized Concept of Watermarking. Watermark Embedding.

TC of Steganography: Generally, one can address steganography types by observing the type of the mask. By other words, the file type of the mask itself is what gives the type of the steganographic system, as further described. **Image Steganography** is the steganographic system $\mathcal{S}$ in which $\mathbf{d} \in \mathcal{D}$ has all the properties of and is represented by **image** type files, regardless of their extension. **Sound/Audio Steganography** is the steganographic system $\mathcal{S}$ in which $\mathbf{d} \in \mathcal{D}$ has all the properties of and is represented by **audio** type files, regardless of their extension. **Video Steganography** is the steganographic system $\mathcal{S}$ in which $\mathbf{d} \in \mathcal{D}$ has all the properties of and is represented by **video** type files, regardless of their extension. **Text Steganography** is the steganographic system $\mathcal{S}$ in which $\mathbf{d} \in \mathcal{D}$ has all the properties of and is represented by **text** type files or simple text passages. **Network Steganography** is the steganographic system $\mathcal{S}$ in which $\mathbf{d} \in \mathcal{D}$ is represented by network traffic elements, such as but not limited to TCP/IP headers.

The cover type must always be designated according to the hands-on applicability of the steganographic system and must always be in accordance with particularities of the overall computer or network system in which the steganographic system is involved.

TC of Watermarking: A categorization of watermarking can be traced based on the visibility of the watermark within the *Watermarked Data*'s representation. When it is meant to be visible and perceptible, we deal with a *visible watermarking*, different in goals and principles from steganography. Where steganography aims for obscurity and secrecy, visible watermarking does the opposite: both the watermark and the protected data it is applied to are intentionally displayed on sight, thus having no secret. On the contrary, when the watermark is meant to be embedded within and withhold some level of secrecy for foreign eyes, the watermarking is known as *invisible* or *transparent*. Adversely from the visible watermarking, the fragile watermarking aims to convert the presence of the watermark. Therefore, not necessarily the watermark itself, but its presence is subject to secrecy, in order to deceive and track unauthorized users of the protected content. *Invisible watermarking* is essentially steganography with application in data ownership protection, as it faces the same methods and shares the assessment aspects.

GO in Steganography: The main goal in steganography is to assure proper data hiding and imperceptibility of the secret's presence. An improper embedding of the secret leads to identifying regions of the mask that seem affected and/or altered, therefore providing details of the secret's presence and positioning, as left side of the following figure illustrates. In the below right image (stego-image), the iris and lower lid region of the baboon contain visibly affected pixels, suggesting that the stego-object is not compliant, signaling and allowing attackers to trace a region in which the secret is hidden. An improper imperceptibility leads to security endangerment, compromising the entire stego-system. A steganographic scheme, therefore, must provide the right balance between the most important vectors of evaluation (treated in Sect. 4), while understanding the weight of imperceptibility and its trade-off (Fig. 4).

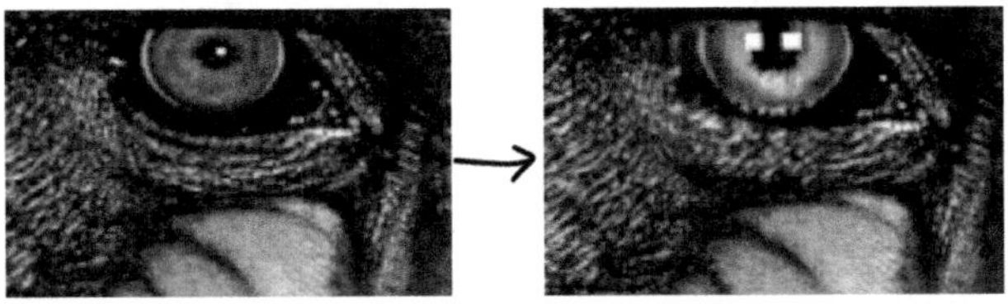

Fig. 4. Failed Image Steganography.

GO in Watermarking: Watermarking is used to protect data's characteristics, such as ownership, distribution tracks, digital rights, authenticity and integrity - aspects that are public and well known. A use case in which watermarking is applied for integrity and authenticity is in medical imaging [4]. A particular case in which secrecy is involved in watermarking is in forensics [2,3], when watermarking's purpose is to track alterations and distributions of protected data without triggering the attacker's attention- fragile watermarking.

4 Vectors of Evaluation (VE). Assessment and Measurement Methods (AMM)

In order to evaluate existing and possible steganographic solution, as for any algorithm, there are certain aspects that aid the evaluation of the existing and proposed solutions. These aspects, *VE*, further referred to in this paper as *VE* represent the evaluation criteria based on which one can asses a specialized method. In the domain of data hiding, there are four such characteristics: robustness, payload capacity, imperceptibility, security [9,10]. **Robustness** aspects asses the ability of the evaluated mechanism to maintain its functionality and performance when faced to erroneous actions (invalid inputs, exceptions, variations etc.) **Payload Capacity** pertains to the maximum lengths and sizes of data the evaluated mechanism is able to process while maintaining its functionality in providing accurate output. **Imperceptibility** measures the amount of data distortion effects that the assessed method is adding to the output it provides. **Security** evaluates how the input data is protected during the processing and manipulation the evaluated method is performing.

VE in Steganography: The robustness aspect imply the ability of a steganosystem to maintain the secret's integrity upon manipulation and attacks performed on the stego-object that it outputs. This aspect is considered by some to be unimportant in steganography [10], while others consider it of great importance. **Payload capacity** implies that between the mask's total feasible capacity (noted as **cap**) and the secret's length (**len(s)**) there must be an inequality relationship, in the sense that $\mathbf{len(s)} \leq \mathbf{cap}$. A stegosystem is considered to be more efficient as its payload capacity increases. A greater payload capacity assures that the secret can be embedded in a single mask instead of it being divided and hid in numerous other masks [12]. **Imperceptibility** addresses the amount of distortion that the mask suffers upon the secret's embedding. A steganographic system fails if the alterations are at least humanly perceptible. **Security**, in steganography, mostly refers to a stegosystem's ability to protect the secret's confidentiality. Although steganography's main goal is to protect the secret inside the mask, there are researchers that tend to forget the importance it holds [13–15], focusing more on VEs such as imperceptibility and payload capacity.

VE in Watermarking: The **robustness** aspect in watermarking addresses the persistence of the watermark upon intentional attacks or unintentional alteration. For instance, in image watermarking, the watermark system should be robust enough to allow the watermark to remain recognisable after repeated paper print - scan actions, or to keep its properties after light adjustments and other manipulations. Here, the deviations parameters a watermark can support are broader than in steganography, due to the fact that human eyes can perceive and identify shapes and images even when altered. Conversely, fragile watermarking aims to provide authentication and verification of digital content, therefore, some studies imply that robustness is not as important here. [16,17]. **Payload capacity** here implies the ability of the watermark to be fully contained within the data it aims to protect- its size associated with a specific decoding error rate. Certain algorithms can have variable payload, [1], while for others [9], this VE is unimportant. **Imperceptibility** is required when there is a need to protect a digital content without significant alterations, with applications in DRM (Digital Rights Management), medical imaging, forensic watermarking, treated in papers like [2–4]. **Security** implies the protection of the watermark against unauthorized alteration or removal. It is relevant in watermarking intended for content tracking, copyright protection and authentication.

Brief Discussion on VE: The same vectors of evaluation measure different aspects in steganography and watermarking, based on each one's intend and purpose. Although fragile watermarking shares common aspects with steganography, it is due to the factual thing that they are closely related, the first being an application of the least. [18,19] When secrecy is brought to watermarking, it fundamentally becomes steganography and therefore the VE converge to purposes and applicability of steganography.

AMM within Steganography: As baseline indicators for data hiding -in general- and for steganography -in particular- are security, imperceptibility, payload capacity and robustness, in practical approach, there is a trade-off between these four VEs which needs to be reduced in order to higher assess a steganographic mechanism. We understand that a form of steganography that can be feasible for nowadays communication means relies on image steganography. As a consequence, the following sections of the present paper will mainly address image steganography.

AMM for Imperceptibility. The imperceptibility often times is reported relatively to the stego-object fidelity resemblance to the mask in its original representation. It represents, therefore, a reference measurement and most of the times it implies non-blind analyse. In the following section, we will provide the mathematical equations for most used of the imperceptibility measurement metrics.

Mean Square Error or MSE, can be calculated for both monochrome (grayscale) and color images. It measures the amount of squared error that happens in the stego-object images. It requires the reference of the original image. For grayscale images, it can be computed based on following equation:

$$MSE = \frac{1}{W * H} * \sum_{x=1}^{W} \sum_{y=1}^{H} (I_{\text{mask}}(x, y) - I_{\text{stego}}(x, y))^2 \tag{1}$$

where W represents the image's width, H represents the image's height, both quantified in terms of pixels number. $I_{\text{mask}}(x, y)$, respectively $I_{\text{stego}}(x, y)$ indicate the pixel values at the position (i,j) inside the mask (cover) and the stego-image.

For color images, it can be computed based on following equation:

$$MSE = \frac{1}{W * H} * \sum_{x=1}^{W} \sum_{y=1}^{H} \sum_{z=1}^{C} (I_{\text{mask}}(x, y, z) - I_{\text{stego}}(x, y, z))^2 \tag{2}$$

where there is an additional implication of the summation for the color channel values (e.g., Red (R), Green (G), Blue (B) in color images using the RGB color model). C represents the overall number of the color channels, z stands for the index of the color channel, $I_{\text{mask}}(x, y, z)$ and $I_{\text{stego}}(x, y, z)$ represent the pixel values at position (x,y) for the z-th color channel. By adding this summation over z, the method accounts for the contribution of each color channel to the overall MSE value. This is important because each color channel denotes different aspects of the image's color information, therefore the MSE calculation should consider the overall color spectrum when evaluating the imperceptibility of a color image. Lower values indicate better imperceptibility and are to look for.

Root Mean Square Error or RMSE, represents a simplified form of MSE. It is derived from MSE by taking the square root of the average squared differences between pixel values in the stego-image and in the cover image, respectively. It can be calculated by

$$RMSE = \sqrt{MSE} \text{ , respectively} \tag{3}$$

$$RMSE = \sqrt{frac1W * H * \sum_{k=1}^{W} \sum_{l=1}^{H} (I_{\text{mask}}(k,l) - I_{\text{stego}}(k,l))^2}, \text{ for grayscale images} \tag{4}$$

$$RMSE = \sqrt{\frac{1}{w * h} * \sum_{k=1}^{w} \sum_{l=1}^{h} \sum_{c=1}^{c} (I_{\text{mask}}(k,l,c) - I_{\text{stego}}(k,l,c))^2}, \text{ for color images.} \tag{5}$$

where k,l represent indices for width (w) and height (h) within the image.

Peak Signal-to-Noise Ratio or PSNR, represents a widely used metric for measuring the imperceptibility in different types of steganography, including image and sound. It requires a reference of the original instance and measures the noise level in the stego-object, expressed in decibels (dB). It relies on MSE and evaluates the quality of reconstructed signals, like the visual fidelity of stego-images compared to their original representations of the masks. PSNR can be calculated based on the following equation

$$PSNR = 10 \cdot \log_{10} \left(\frac{\text{MAX}^2}{\text{MSE}} \right) \tag{6}$$

where MAX is the maximum possible pixel value of the image and MSE is the Mean Square Error between the cover and stego-images, computed accordingly for grayscale and color images, as treated in Sect. 4.

PSNR values are interpred as follows: PSNR $>$ 30 dB- Excellent quality (imperceptible difference), 20 dB $<$ PSNR $<$ 30 dB- Good quality (slightly perceptible difference), PSNR $<$ 20 dB- Poor quality (highly perceptible difference).

Structural Similarity Index, SSIM or SSI, is metric used to evaluates the similarity between the stego-image and its original mask. SSI is used to quantify the structural information and the luminance distortion that occures between the stego and original image. SSI contrasts the luminance, contract and brightness of corresponding areas within the stego and original cover image, while considering local variations in texture and intensity and spatial relationship between pixels.

By other words, one can understand the components SSI is calculated based on: **Luminance Comparison**- measures the similiarity of pixel intensities for corresponding image locations, **Contrast Comparison**- measures the dissimilarities in contrast between corresponding areas within the stego-image and cover, **Structure Comparison**- measures the structural similarity between corresponding areas within the stego and cover images. Values of SSI are interpreted as follows: SSIM = 1 denotes perfect similarity between original cover and stego-image, SSIM - -1, rare cases, denotes perfect dissimilarity between cover and stego-image, SSIM - 0 denotes no similarity between the cover and stego-image. Values closer to 1 represent better a better quallity of SSIM. SSIM can be calculated based on the following:

$$\mathrm{SSIM(cover, stego)} = \frac{(2\mu_{\mathrm{cover}}\mu_{\mathrm{stego}} + C_1)(2\sigma_{\mathrm{cover,stego}} + C_2)}{(\mu^2_{\mathrm{cover}} + \mu^2_{\mathrm{stego}} + C_1)(\sigma^2_{\mathrm{cover}} + \sigma^2_{\mathrm{stego}} + C_2)} \tag{7}$$

where μ_{cover} and μ_{stego} are the mean values of intensity of luminance of the cover and stego images, respectively, $\sigma^2_{\mathrm{cover}}$ and $\sigma^2_{\mathrm{stego}}$ are the standard deviation of signal contrast of the cover and stego images, respectively, C_1 and C_2 are constants added to prevent division by zero and therefore stabilize the denominator with

$$C_1 = (k_1 L)^2, C_2 = (k_2 L)^2 \tag{8}$$

where k_1 and k_2 represent constants chosen to be small positive values, often set to 0.01 or 0.03, and L is the dynamic range of pixel values in the stego-image and cover. SSIM considers both visual fidelity and structural integrity, therefore it can also be used as an assesment metric for robustness.

Mean Structural Similarity Index (MSSIM) represents a form of SSIM that addresses the images in their whole or multiple areas and not only singular patches, like SSIM does. Test [21] have demonstrated that MSSIM outperforms SSIM alone. However, the said results are not specific for steganography but for different types of image processing. It assesses the imperceptibility based on images' structural content, as quantified by the SSIM metric and is computed based on the following equation:

$$\mathrm{MSSIM(cover, stego)} = \frac{1}{N}\sum_{i=1}^{N} \mathrm{SSIM}(p_i^{\mathrm{cover}}, p_i^{\mathrm{stego}}) \tag{9}$$

where N represents the total number of windows (areas) considered in the images, p_i^{cover} and p_i^{stego} denote the i-th index region in the cover and stego image, respectively, and $\mathrm{SSIM}(p_i^{\mathrm{cover}}, p_i^{\mathrm{stego}}$ represent the SSIM value calculated for the patch in cover and stego at the i-th index.

Multi-Scale Structural Similarity Index or MS-SSIM is a multi-scale extension of SSIM that addresses the structural similarities in images at multiple scales. MS-SSIM is particularly of help in those scenarios in which images contain structures of different scales (e.g. medical images). MS-SSIM metric addresses to a reference and a stego-image and involves decomposing the images into multiple scales based either on a Gaussian pyramid or similar multi-scale decomposition techniques. A simplified equation of MS-SSIM is the following:

$$\text{MS-SSIM}(\text{cover}, \text{stego}) = \left[\prod_{i=1}^{N} \text{SSIM}_i(\text{cover}, \text{stego})\right]^{\frac{1}{N}} \quad (10)$$

where $\text{SSIM}_i(\text{cover}, \text{stego}$ represent the SSIM value for each i-th scale and N represnets the total number of considered scales. In steganography research, MS-SSIM is not as popular and used as SSIM and MSSIM, due to its need of further developing in order to become more stable.

Universal Image Quality Index, UIQI or QI is yet another metric that applies to a stego-image with reference of the original cover. Although SSIM is considered to be a more advanced and sophisticated method, there are studies that address the usage of QI metric [22,25,26]. QI addresses the charachteristics of human visual system and accounts the luminnce, contrast and structure of images. It can be calculated based on the following:

$$\text{UIQI} = \frac{4\sigma_{xy}\mu_x\mu_y}{(\sigma_x^2 + \sigma_y^2)(\mu_x^2 + \mu_y^2)} \quad (11)$$

where σ_{xy} is the covariance between the stego and reference and μ_x and μ_y represent the means of pixel values in the stego-image and cover. A value closer to 1 represent a better score for UIQI, implying a higher imperceptibility.

Feature Similarity Index or FSIM, is yet another metric that addresses the characteristics of the human visual system and measures how well the stego-image preserves the visual similarity with the mask image.

The FSI metric is based structural elements. A high FSI value indicates a closer resamblence between the two images, while lower FSI values indicate significant dissimilarities and alterations to the visual context of the stego-image. It represents an important tool in steganalysis and steganography research [6,7] and was firstly published by [8]. The FSIM can be computed by the following simplified equation:

$$FSIM = \frac{\sum x \in \Omega L_s * PC_m(x)}{\sum x \in \Omega PC_m(x)}, \text{ for grayscale images} \quad (12)$$

where Ω represents the entire image, L_s represents the brightness correlation between the images considered, x represents the current position marker, as also illustrated in paper [24].

In order to be possible to be applied for color images, a conversion is needed between RGB to YIQ images.

Kullback-Leibler Divergence, KL Divergence or relative entropy is a metric that evaluates how a probability distribution diverges from a second expected probability distribution. It is asymmetric, meaning a difference of one probability distribution and another is not necessarily the same as the difference between the second distribution and the first. Lower values of KL Divergence denote that the stego-image statistical properties closely resemble those of the cover, indicating a higher efficiency of the steganographic algorithm. This metric has been used in studies like [20]. The KL Divergence can be calculated based on the following forumula:

$$D_{\mathrm{KL}}(cover||stego) = \sum_{x \in X} cover(x) \log \left(\frac{cover(x)}{stego(x)} \right) \quad (13)$$

where $D_{\mathrm{KL}}(\mathrm{cover}|\mathrm{stego})$ is the KL Divergence that measures how much the stego distribution diverges from the cover distribution, *cover(x)* and *stego(x)* are the probability distribution of the items being compared, cover image and stego-image, respectively, and x represents the possible outcome of events, depending on the color space and representation used.

Visual Information Fidelity or VIF, is a metric used to quantify the fidelity of visual information of the stego-image reported to the original cover. By other words, VIF evaluates how well the steganographic algorithm maintained the perceptual similarity between the original cover and the image containing embedded information. Higher values pf VIF indicate a better quality of the image. VIF values can be calculated as the ratio of the sum of mutual information terms over all subbands in the stego image to the sum of mutual information terms of all subbands in the stego-image and is broadly explained in [5].

BRISQUE Score or Blind/Referenceless Image Spatial Quality Evaluator, is a metric that does not require the original image as a reference, thus the terms *Blind* or *Referenceless*. It is used to quantify the naturalness or perceptual quality of an image by analyzing its spatial domain statistics. Higher BRISQUE scores typically denote a poorer quality of the image or a less natural-looking representation. Lower BRISQUE scores are to look for in steganography, as also mentioned by researches in papers like [21,23].

Mean Opinion Score or MOS, is a subjective metric mostly used to evaluate the humanly perceived quality of audio and video data. It relies on presenting observes with data samples and asking them to rate the quality they perceive each sample has, on a predefined scale, commonly from 1 to 5 or 1 to 10. Higher values indicate higher quality levels. The MOS value is then typically obtained by averaging the sums of the provided individual scores. Study [27] involves 10 human observers, 10 men and 10 women in assessing a MOS score.

AMM for Payload Capacity: Bits per Pixel or BPP, is a metric that calculates the efficiency of a steganographic algorithm in terms of capacity. It addresses the amount of secret data that can be embedded within each pixel of the cover. For grayscale images, one pixel is composed of a singular byte, while per color images there are either three (or even four - including the alpha channel) bytes per pixel. BPP is always an important factor for each steganographic algorithm, as it represents one of the features it offers. It aids in algorithm comparison and balancing the capacity payload with the imperceptibility VEs. It has been treated in studies like [33,34] as the term Bits per Channel or Bits per Byte. It can be calculated based on the following equation:

$$\mathrm{bpp} = \frac{\text{Total bits for embedding}}{\text{H*W*C}} \quad (14)$$

where H and W represent the number of pixels on the height and the width, respectively, in the cover image, and C represents the number of channels for each pixel.

Bit count per Non-Zero DCT Coefficient, bpnz or Bpc, is a metric used to calculate the average number of data bits that can be embedded within each non-zero Discrete Cosine Transform (DCT) coefficient in the cover image. It is broadly used in JPEG steganography. It aids in payload capacity assessment, research and algorithm comparison, optimization consideration. Higher bpnz values while minimizing perceptual alterations are to look for. It can be calculated based on the following equation:

$$\text{bpnz} = \frac{\text{Total bits for embedding}}{\text{Total number of non-zero DCT coefficients}} \tag{15}$$

AMM for Security- Steganalysis: Security in steganography is often times assessed by steganalysis techniques. Steganalysis is the science and study of steganography methods and results with the purpose to provide conclusions regarding the presence and content of a secret within a mask. A categorization of steganalysis based on its methodology is represented by: **Specific** steganalysis - when the metrics and tools are dedicated for the evaluation and identification of a specific steganographic method or technique and **Universal** steganalysis -when the metrics and tools are dedicated to identify the presence and content of a secret regardless of the involved steganographic method or technique.

Regular Singular Analysis or RS, is a steganalysis method with the purpose to assert the presence of a secret message inside an image. RS quantifies the amount of noise inside a tested image, by dividing the image in sections (groups), measuring the amount of noise present in each section (using the function f) and comparing them to the measured noise when LSB is simulated with function F and F_1, respectively. More noise results in larger values of the discrimination function *f*. **Regular groups (R)**: are those sections in which the LSB alterations conducted to an increased amount of noise. **Singular groups (S)**: are those sections in which the LSB alterations decreased the amount of noise. **Unusable goups (U)**: are those sections in which LSB alterations did not increased nor did they decreased the level of noise. A mask would normally produce more R groups.

Subtractive Pixel Adjacency Matrix or SPAM, is a method proposed in [35]. SPAM performs well in aserting the presence of an embedded message in case of Spatial Domain steganography. [36] It considers the difference between adjacent pixels and represents yet another analysis method for LSB steganography, but has proven to be efficient for DCT also, therefore classifying it as an universal steganographic technique [35]. SPAM implies the calculation of a matrix of differences between intensity values of pixels on adjacent positions. For each pixel, taking into account a neighboring structure (Moore, Von Neumann), the difference between its values and its neighbors is calculated and accommodated into the matrix [35].

Textural Features based Universal Steganalysis was proposed as a set of features for universal steganalysis based on an cover's normalized histograms of the local linear transform coefficients by using the Fisher Linear Discriminant (FLD). This method applies to grayscale images and is also reported to be applicable to JPEG and JPEG2000 steganography.

Spatial Rich Model Features or SRM, is an effective steganalysis technique that asserts the presence of an embedded message based on statistical characteristics of neighboring residual noise samples [37]. SRM was proposed by the study [41] and is implemented using machine learning - ensemble classifier - and trained on masks and stego-images.

Universal JPEG Steganalysis As it is not specialized in a singular algorithm, techniques in this domain address the statistical features of JPEG images that are typically affected by data embedding process. This implies that methods of this manner trace conclusions about noise petterns, distortion and anomalies that the steganographic process carry upon embedding. Some of the research works in Universal JPEG Steganalysis spectrum are [38,39] that take into consideration DCT features, [39] - a study that proposes a method of JPEG Steganalysis based on collaborative representation and [40] - a study that addresses the impact in JPEG Steganalysis of different software techniques based on transfer learning.

Deep Learning Methods have emerged the image steganlysis spectrum due to their capability of automated learning of discriminative features from significantly large image datasets. Convolutional Neural Networks, or CNNs, are often times implied in stagnalysis due to their ability to effectively capture spatial patterns and hierarchical features in images. CNNs require references in learning, therefore are trained on datasets containing both the stego-image and the masks. CNNs results are of greater accuracy than SPAM methods, but do not exceed SRM. Paper [40] compares several deep learning methods in steganography, such as EfficientNet, MixNet [43], DenseNet, but there are many other methods, like XuNet, YeNet, QianNEt, ZhuNet [44] that performed well in image steganalysis.

AMM for Robustness: Bit Error Ratio or BER, is used to measure the integrity of data obtained after extraction from stego-image, as the ratio of altered bits to the total number of initially embedded bits. It is useful in error correction, steganographic research and data integrity assessment. It is typically expressed as percentage with lower values being preferred. Higher values of BER denote poorer accuracy in the extracted data, when the Bit Error Ratio is expressed as follows. BER can be calculated based on the following equation:

$$\mathrm{BER} = \frac{\mathrm{E}}{\mathrm{L}} \times 100\% \tag{16}$$

where E is the total number of incorrectly decoded bits and L represents the lenght expressed in bits of the embedded data. Some of the steganographic researches that addresses the BER metric are [19,42,45–50].

Error Rate Extraction or R_e, is a metric similar to BER, with the difference that R_e is not expressed as percentile figures. It tests the secret extraction, therefore is useful in assessing the robustness character of a steganographic method. A R_e value of 0 indicates a 100% accurate and intact message upon extraction. Higher values are inversely proportional to the accuracy and integrity of the extracted data. Authors that addressed R_e in their steganographic work are those of papers like [51–57].

Other AMMs: The above sections consists of assessment methods we conclude to be mostly suitable for steganography. However, the list is not exhaustive. 2D-Correlation Coefficient (CC) is another metric used to asses the coefficient of the correlation between the embedded and extracted data. The Character Error Ratio, or CER, represents a metric method used to assess the quantity of error occurred in data extraction of text messages. [18,19] PSNR, MSE, SSIM, MSSIM metrics are also used for the assessment of robustenss in steganography [42,45,48,58]. VIF [5], QI Index [22,25,26], FSIM [6–8], Pearson Correlation oefficient are used in studies like [20], Normalized Cross Coefficient are treated in papers like [27–32]. These correlation coefficient-based metrics are more adequate for measuring the proccess of message extraction [24].

5 Steganography Contrasted with Watermarking

Below is a tabular representation containing clear key aspects that are either similar or dissimilar between the two: steganography and watermarking. Traits as objective, message's characteristic, cover's purpose, rights of ownership upon cover, purpose, reversibility and failure scenarios are being depicted (Table 1).

Table 1. Tabular Comparison: Steganography vs. Watermarking

Trait	Steganography	Watermarking
Objective	Protection of the message	Protection of the cover media
Message	Is secret	(The watermark) is public
Cover's Purpose	Mask	Protected media
Supported Types for Cover (mainly)	Any: from text, alphabetical letters to digital	Digital/Physical data
Rights upon Cover	Not required	Required
Intended for	Data at rest and in transit	Data in transit
Recovery Method	Blind (original data is not needed)	Either blind or non-blind (original data is needed)
Accepted Failure Rate for Recovery	None. The secret's integrity must be exact upon recovery	Enough for the watermark to be recognizable
Main Assessment Aspects	Security, payload capacity, imperceptibility	Robustness, security
Apparentness	Invisible	Visible and invisible- for fragile
Key Presence	Not required	Not required
Purpose	Confidentiality, censorship evasion, covert operations, secure storage	Ownership, authentication, copy control
Reversibility	Reversible, irreversible (on embedding content)	Reversible, irreversible (on embedded content)
Domain of Embedding	Spatial, Transform, File Structure (ex. Metadata), Network Protocols	Spatial, Transform, File Structure (ex. Copyright Metadata)
Failure Scenario	Secret is detectable and/or extracted	Watermark is removed and/or replaced

6 Summary

Steganography alongside Watermarking are, by no doubt, a subject of interest for researchers. Along the relatively recent years (aproximatively three decades), there have been advances in what regards the evaluation of steganographic and watermarking techniques with the ultimate purpose to enhance steganography as a main domain under Security and Data Protection, its methodology and applicability. Steganography's interoperability and availability to integrate adaptive and learning techniques within its development and analysis practices illustrates how vast and robust this domain in fact is. Although in the past steganography and watermarking were not so divergent as sciences, today it is important to understand where these two fields converge and what are their aspects that makes them, in fact, so different. Invisible watermarking, in its similarities of practices and principles with steganography, represents, essentially, an application of steganography in the context of data ownership retribution and rights protection.

Steganography can be categorized based on its cover type and further more categorized based on the cover's data types and formats. Therefore, it is a wide area to research and comprehend. There are dilemmas that universally raise for steganography and further more specialized dilemmas for certain types of steganography. Conversely, watermarking is categorized based on perceptibility (visible or invisible), purpose (ownership, authentication, copy control) or application area (watermarking of image, audio, text, video).

Analysis performed on stego-systems can be specialized - when it addresses on the identification of a certain steganographic algorithm, and universal - when it is portable to the detection and perceiving of many. Steganogrphy can be irreversible or reversible, with particular aspects to be taken into consideration for each type. Reversible steganography might lead to providing fingerprints in case of steganalysis. Steganalysis methods might require a reference for benchmarking or can be blindly applied. Similar to steganography, watermarking can be irreversible (permanent, non-recoverable watermarking) or reversible (lossless, recoverable watermarking). The characteristics of reversibility aspects in steganography and watermarking address different components for each. For the first, it is the mask retrieval that dictates the reversibility characteristic, while for the second it is the possibility to retrieve the added watermark that assess whether watermarking is reversible.

Despite that the sciences of steganography and watermarking have been at some point in history not so different, the natural evolution has separated them. Both sciences regard data embedding and hiding at some level, but not both demand secrecy as a core trait. Nonetheless, secrecy is involved at times-in watermarking and always in steganography. When watermarking demands secrecy, it is essentially steganography. Researchers can use all these conclusions in order to develop their methods accordingly and further and valuable develop the domain.

References

1. Zhao, L., Qian, Z., Qin, C., Xie, Y.: An image self-recovery approach with variable payload. In: 2014 9th IEEE Conference on Industrial Electronics and Applications, pp. 1254–1257, Hangzhou, China (2014). https://doi.org/10.1109/ICIEA.2014.6931360
2. Zhou, G., Lv, D.: An overview of digital watermarking in image forensics. In: 2011 Fourth International Joint Conference on Computational Sciences and Optimization, pp. 332–335, Kunming and Lijiang City, China (2011). https://doi.org/10.1109/CSO.2011.85
3. Elavarasan, S.R., Florence, S.M.: Forensic video watermarking. In: 2023 2nd International Conference on Automation, Computing and Renewable Systems (ICACRS), pp. 1838–1840, Pudukkottai, India (2023). https://doi.org/10.1109/ICACRS58579.2023.10404733
4. Mestiri, A., Kricha, A., Sakly, A., Mtibaa, A.: Watermarking for integrity, authentication and security of medical imaging. In: 2017 14th International Multi-Conference on Systems, Signals & Devices (SSD), pp. 475–481, Marrakech, Morocco (2017). https://doi.org/10.1109/SSD.2017.8166967
5. Sheikh, H.R., Bovik, A.C.: Image information and visual quality. IEEE Trans. Image Process. **15**(2), 430–444 (2006). https://doi.org/10.1109/TIP.2005.859378
6. Subhedar, M., Mankar, V.: Secure image steganography using framelet transform and bidiagonal SVD. Multimedia Tools Appl. **79** (2020). https://doi.org/10.1007/s11042-019-08221-9
7. Gaurav, K., Ghanekar, U.: Image steganography based on canny edge detection, dilation operator and hybrid coding. J. Inform. Secur. Appl. **41**, 41–51 (2018). https://doi.org/10.1016/j.jisa.2018.05.001
8. Zhang, L., Zhang, L., Mou, X., Zhang, D.: FSIM: a feature similarity index for image quality assessment. IEEE Trans. Image Process. **20**, 2378–2386 (2011). https://doi.org/10.2307/1577889
9. Cheddad, A., Condell, J., Curran, K., McKevitt, P.: Digital image steganography: survey and analysis of current methods. Signal Process. (2010). https://doi.org/10.1016/j.sigpro.2009.08.010
10. Fridrich, J.: Steganography in digital media: principles, algorithms, and applications. IEEE Signal Process. Mag. **28**, 142–144 (2011). https://doi.org/10.1109/msp.2011.941841
11. Majeed, M.A., Sulaiman, R., Shukur, Z., Hasan, M.K.: A review on text steganography techniques. Mathematics **9** (2021). https://doi.org/10.3390/math9212829
12. Setiadi, D.R.I.: PSNR vs SSIM: imperceptibility quality assessment for image steganography. Multimedia Tools Appl. **80**, 1–22 (2021). https://doi.org/10.1007/s11042-020-10035-z
13. Bai, J., Chang, C.C., Nguyen, T.S., Zhu, C., Liu, Y.: A high payload steganographic algorithm based on edge detection. Displays **46**, 42–51 (2017). https://doi.org/10.1016/j.displa.2016.12.004
14. Wang, Y., et al.: High-capacity adaptive steganography based on LSB and hamming code. Optik **213**, 164685 (2020). https://doi.org/10.1016/j.ijleo.2020.164685
15. Ghosal, S.K., Mandal, J.K., Sarkar, R.: High payload image steganography based on Laplacian of Gaussian (LoG) edge detector. Multimedia Tools Appl. **77**(23), 30403–30418 (2018). https://doi.org/10.1007/s11042-018-6126-y
16. Zhang, Z., Sun, H., Gao, S., Jin, S.: Self-recovery reversible image watermarking algorithm. PLoS One **13**, 1–27 (2018). https://doi.org/10.1371/journal.pone.0199143

17. Kim, C., Yang, C.N.: Self-embedding fragile watermarking scheme to detect image tampering using AMBTC and OPAP approaches. Appl. Sci. **11**, 1–21 (2021). https://doi.org/10.3390/app11031146
18. Jayapandiyan, J.R., Kavita, C., Sakthivel, K.: Enhanced least significant bit replacement algorithm in spatial domain of steganography using character sequence optimization. IEEE Access (2020). https://doi.org/10.1109/ACCESS.2020.3009234
19. Setiadi, D.R.I.M.: Improved payload capacity in LSB image steganography uses dilated hybrid edge detection. J. King Saud Univ. Comput. Inform. Sci. **34**, 104–114 (2022). https://doi.org/10.1016/J.JKSUCI.2019.12.007
20. Sabeti, V., Sobhani, M., Hasheminejad, S.M.H.: An adaptive image steganography method based on integer wavelet transform using genetic algorithm. Comput. Electr. Eng. **99**, 107809 (2022). https://doi.org/10.1016/J.COMPELECENG.2022.107809
21. Mittal, A., Moorthy, A.K., Bovik, A.C.: Blind/referenceless image spatial quality evaluator. In: Conference Record of the Asilomar Conference on Signals, Systems and Computers, pp. 723–727 (2011). https://doi.org/10.1109/ACSSC.2011.6190099
22. Wang, Z., Bovik, A.C.: A universal image quality index. IEEE Sig. Process. Lett. **9**, 81–84 (2002). https://doi.org/10.1109/97.995823
23. Wu, H.T., Tang, S., Huang, J., Shi, Y.Q.: A novel reversible data hiding method with image contrast enhancement. Sig. Process. Image Commun. **62**, 64–73 (2018). https://doi.org/10.1016/J.IMAGE.2017.12.006
24. Setiadi, D.R.I.M., Rustad, S., Andono, P.N., Shidik, G.F.: Digital image steganography survey and investigation (goal, assessment, method, development, and dataset). Sig. Process. **206**, 108908 (2023). https://doi.org/10.1016/j.sigpro.2022.108908
25. Hameed, M.A., Hassaballah, M., Aly, S., Awad, A.I.: An adaptive image steganography method based on histogram of oriented gradient and PVD-LSB techniques. IEEE Access **7**, 185189–185204 (2019). https://doi.org/10.1109/ACCESS.2019.2960254
26. Sahu, A.K., Swain, G., Babu, E.S.: Digital image steganography using bit flipping. Cybern. Inform. Technol. **18**, 69–80 (2018). https://doi.org/10.2478/cait-2018-0006
27. Muhammad, K., Sajjad, M., Mehmood, I., Rho, S., Baik, S.W.: Image steganography using uncorrelated color space and its application for security of visual contents in online social networks. Future Gener. Comput. Syst. **86**, 951–960 (2018). https://doi.org/10.1016/j.future.2016.11.029
28. Thanki, R., Borra, S.: A color image steganography in hybrid FRT-DWT domain. J. Inform. Secur. Appl. **40**, 92–102 (2018). https://doi.org/10.1016/J.JISA.2018.03.004
29. Mandal, P.C., Mukherjee, I., Chatterji, B.N.: High capacity reversible and secured data hiding in images using interpolation and difference expansion technique. Multimedia Tools Appl. **80**, 3623–3644 (2021). https://doi.org/10.1007/s11042-020-09341-3
30. Mehta, D., Bhatti, D.: Blind image steganography algorithm development which resists JPEG compression attack. Multimedia Tools Appl. **81**, 459–479 (2022). https://doi.org/10.1007/s11042-021-11351-8
31. Datta, B., Roy, S., Roy, S., Bandyopadhyay, S.K.: Multi-bit robust image steganography based on modular arithmetic. Multimedia Tools Appl. **78**, 1511–1546 (2019). https://doi.org/10.1007/s11042-018-6195-y

32. Mukherjee, N., Paul, G., Saha, S.K., Burman, D.: A PVD based high capacity steganography algorithm with embedding in non-sequential position. Multimedia Tools Appl. 1–31 (2020). https://doi.org/10.1007/s11042-019-08178-9
33. Swain, G.: Digital image steganography using eight-directional PVD against RS analysis and PDH analysis. Adv. Multimedia 2018 (2018). https://doi.org/10.1155/2018/4847098
34. Liao, X., Yu, Y., Li, B., Li, Z., Qin, Z.: A new payload partition strategy in color image steganography. IEEE Trans. Circ. Syst. Video Technol. **30**, 685–696 (2020). https://doi.org/10.1109/TCSVT.2019.2896270
35. Pevný, T., Bas, P., Fridrich, J.: Steganalysis by subtractive pixel adjacency matrix. IEEE Trans. Inform. Forensics Secur. **5**, 215–224 (2010). https://doi.org/10.1007/s11432-013-4793-x
36. Gu, X., Guo, J.: A study on subtractive pixel adjacency matrix features. Multimedia Tools Appl. **78**(14), 19681–19695 (2019). https://doi.org/10.1007/s11042-019-7285-1
37. Wang, P., Wei, Z., Xiao, L.: Pure spatial rich model features for digital image steganalysis. Multimedia Tools Appl. **75**(5), 2897–2912 (2015). https://doi.org/10.1007/s11042-015-2521-9
38. Khalilollahi, S.M.S., Mansouri, A.: JPEG steganalysis using the relations between DCT coefficients. In: 2022 International Conference on Machine Vision and Image Processing (MVIP), pp. 1–4, Ahvaz, Iran (2022). https://doi.org/10.1109/MVIP53647.2022.9738785
39. Guo, J., Guo, Y., Li, L., Li, M.: A universal JPEG image steganalysis method based on collaborative representation. In: Proceedings of the 2014 IEEE International Conference on Security, Pattern Analysis, and Cybernetics (SPAC), pp. 285–289 , Wuhan, China (2014). https://doi.org/10.1109/SPAC.2014.6982700
40. Padmasiri, A.T., Hettiarachchi, S.: Impact on JPEG image steganalysis using transfer learning. In: 2021 10th International Conference on Information and Automation for Sustainability (ICIAfS), pp. 234–239, Negambo, Sri Lanka (2021). https://doi.org/10.1109/ICIAfS52090.2021.9605924
41. Fridrich, J., Kodovsky, J.: Rich models for steganalysis of digital images. IEEE Trans. Inform. Forensics Secur. **7**, 868–882 (2012). https://doi.org/10.1109/TIFS.2012.2190402
42. Kadhim, I.J., Premaratne, P., Vial, P.J.: High capacity adaptive image steganography with cover region selection using dual-tree complex wavelet transform. Cogn. Syst. Res. **60**, 20–32 (2020). https://doi.org/10.1016/j.cogsys.2019.11.002
43. Amrutha, E., Arivazhagan, S., Jebarani, W.S.L.: MixNet: a robust mixture of convolutional neural networks as feature extractors to detect stego images created by content-adaptive steganography. Neural Process. Lett. **54**, 853–870 (2022). https://doi.org/10.1007/s11063-021-10661-0
44. Zhang, R., Zhu, F., Liu, J., Liu, G.: Efficient feature learning and multi-size image steganalysis based on CNN. arXiv preprint, arXiv:1807.11428 (2018). http://arxiv.org/abs/1807.11428
45. Sahu, A.K., Swain, G., Sahu, M., Hemalatha, J.: Multi-directional block based PVD and modulus function image steganography to avoid FOBP and IEP. J. Inform. Secur. Appl. **58**, 102808 (2021). https://doi.org/10.1016/j.jisa.2021.102808
46. Sahu, A.K., Swain, G.: High fidelity based reversible data hiding using modified LSB matching and pixel difference. J. King Saud Univ. Comput. Inform. Sci. **34**, 1395–1409 (2022). https://doi.org/10.1016/J.JKSUCI.2019.07.004

47. Rustad, S., Setiadi, D.R.I.M., Syukur, A., Andono, P.N.: Inverted LSB image steganography using adaptive pattern to improve imperceptibility. J. King Saud Univ. Comput. Inform. Sci. **34**, 3559–3568 (2022). https://doi.org/10.1016/J.JKSUCI.2020.12.017
48. Sukumar, A., et al.: Robust image steganography approach based on RIWT-laplacian pyramid and histogram shifting using deep learning. Multimedia Syst. **27**(6), 651–666 (2021). https://doi.org/10.1007/s00530-020-00665-6
49. Di, F., Zhang, M., Liao, X., Liu, J.: High-fidelity reversible data hiding by quadtree-based pixel value ordering. Multimedia Tools Appl. **78**, 7125–7141 (2019). https://doi.org/10.1007/s11042-018-6469-4
50. Kadhim, I.J., Premaratne, P., Vial, P.J.: Improved image steganography based on super-pixel and coefficient-plane-selection. Signal Process. **171**, 107481 (2020). https://doi.org/10.1016/j.sigpro.2020.107481
51. Tao, J., Li, S., Zhang, X., Wang, Z.: Towards robust image steganography. IEEE Trans. Circ. Syst. Video Technol. **29**, 594–600 (2019). https://doi.org/10.1109/TCSVT.2018.2881118
52. Zhu, Z., Zheng, N., Qiao, T., Xu, M.: Robust steganography by modifying sign of DCT coefficients. IEEE Access **7**, 168613–168628 (2019). https://doi.org/10.1109/ACCESS.2019.2953504
53. Lu, W., Zhang, J., Zhao, X., Zhang, W., Huang, J.: Secure robust JPEG steganography based on autoEncoder with adaptive BCH encoding. IEEE Trans. Circ. Syst. Video Technol. **31**, 2909–2922 (2021). https://doi.org/10.1109/TCSVT.2020.3027843
54. Cogranne, R., Giboulot, Q., Bas, P.: Steganography by minimizing statistical detectability: the cases of JPEG and color images. In: Proceedings of the 2020 ACM Workshop on Information Hiding and Multimedia Security (IH MMSec), pp. 161–167 (2020). https://doi.org/10.1145/3369412.3395075
55. Zhang, Y., et al.: On the fault-tolerant performance for a class of robust image steganography. Signal Process. **146**, 99–111 (2018). https://doi.org/10.1016/j.sigpro.2018.01.011
56. Zhang, Y., et al.: Dither modulation based adaptive steganography resisting JPEG compression and statistic detection. Multimedia Tools Appl. **77**(17), 17913–17935 (2018). https://doi.org/10.1007/s11042-017-4506-3
57. Kusuma, E.J., Sari, C.A., Rachmawanto, E.H., Setiadi, D.: A combination of inverted LSB, RSA, and arnold transformation to get secure and imperceptible image steganography. J. ICT Res. Appl. **12**, 103 (2018). https://doi.org/10.5614/itbj.ict.res.appl.2018.12.2.1
58. Sahu, A.K., Swain, G.: A novel n-rightmost bit replacement image steganography technique. 3D Res. **10**, 1–18 (2019). https://doi.org/10.1007/s13319-018-0211-x
59. Elavarasan, S.R., Florence, S.M.: Forensic video watermarking. In: ICACRS 2023, pp. 1838–1840, India (2023)

Enhanced Devices Security Using Elliptic Curves Defined over Particular Spaces

Oana-Adriana Ticleanu(✉), Daniel Hunyadi, and Nicolae Constantinescu

Department of Mathematics and Informatics, Lucian Blaga University of Sibiu, Sibiu, Romania
{oana.ticleanu,daniel.hunyadi,nicolae.constantinescu}@ulbsibiu.ro

Abstract. Elliptic curve cryptography (ECC) has become a central pillar of modern security due to its efficiency and high level of security provided per bit. However, traditional implementations face challenges in resource-constrained environments, such as IoT devices and wireless sensor networks, due to high computational demands and vulnerabilities to advanced attacks. This paper proposes an innovative extension of ECC, using elliptic curves defined on particular spaces to address these limitations . By generalizing the classical elliptic curve equation with a space-dependent function $f(x, y)$, we optimize the density of valid points and improve computational efficiency, while maintaining security properties. Experimental results show a 15% reduction in execution time for scalar multiplication and a 12% decrease in energy consumption compared to traditional ECC. The proposed method also demonstrates increased resistance to brute force and side-channel attacks, due to the complexity introduced by the function $f(x, y)$. Scalability tests confirm the applicability of the solution in large networks, making it suitable for industrial and critical infrastructures. This work provides a solid theoretical basis and practical validation for the integration of particular spaces in ECC, opening new perspectives for efficient and secure cryptographic solutions. Future research directions include exploring other types of particular spaces, hybridizing with post-quantum cryptographic schemes, and testing the methodology in blockchain and other distributed systems.

Keywords: Cryptography over elliptic curves · Scalar multiplication · EC space reduction

1 Introduction

In an increasingly connected technological landscape, device security has become a central topic of modern research. The emergence and rapid development of ecosystems based on the Internet of Things (IoT), wireless sensor networks, and cloud infrastructures have radically transformed the way we manage data and digital communication. However, this technological expansion also implies a

D. Simian and L. F. Stoica (Eds.): MDIS 2024, CCIS 2486, pp. 239–255, 2025.
https://doi.org/10.1007/978-3-031-87386-7_16

significant increase in security risks. IoT devices are often vulnerable to cyber-attacks due to their limited hardware and software resources, as well as constant network connectivity. In this context, ensuring the security of these devices becomes an essential challenge.

Elliptic curve cryptography (ECC) is one of the most promising solutions to this challenge. Using the mathematical properties of elliptic curves, ECC offers a high level of security using much smaller cryptographic keys compared to traditional methods, such as RSA. This makes ECC ideal for devices with limited resources, where optimizing memory and power consumption is essential. However, although ECC has proven effective in many applications, there are important limitations. Traditional ECC algorithms are computationally expensive on devices with limited processing power and are vulnerable to advanced attacks, such as those based on temporal analysis or side-channel attacks. Furthermore, the development of quantum computers threatens the robustness of traditional ECC, calling into question its long-term use.

An emerging field in cryptography is the use of elliptic curves defined on particular spaces. These spaces, such as hyperbolic surfaces or special geometric manifolds, offer new opportunities to optimize the properties of elliptic curves. By adapting the geometric characteristics of the curves, it is possible to create more efficient cryptographic schemes that are more resistant to advanced attacks. However, research in this direction is still in its infancy, and the potential of these structures has not been fully exploited.

The present paper addresses this gap by proposing a new theoretical and methodological framework for the use of elliptic curves defined on particular spaces. The main goal is to improve the efficiency and security of elliptic curve-based cryptography, especially for devices with limited resources. By optimizing curve geometry and adapting cryptographic algorithms, this approach can reduce resource consumption, increase resistance to attacks, and prepare digital infrastructures for emerging threats, such as attacks based on quantum computers. Elliptic curve cryptography (ECC) has become a mainstream solution in the field of cybersecurity due to its advantageous ratio between the level of security provided and the resources used [19,20]. It is applicable in a variety of environments, such as IoT networks and critical systems, where resource efficiency is crucial [21]. However, these implementations face limitations in resource-constrained environments.

There are ongoing studies on several levels, related to streamlining the mathematical models used, as well as how cryptographic operations on elliptic curves defined on particular spaces can be optimized to increase efficiency and security in constrained environments.

Although recent studies have improved the efficiency of ECC [22,23], the use of curves defined on particular spaces to overcome current limitations is underexplored. This offers an opportunity to address vulnerabilities to attacks and reduce the complexity of cryptographic operations.

This paper explores the following research questions: (1) How can particular spaces improve the density of valid points on elliptic curves to optimize efficiency? (2) What algorithms can be developed to leverage the geometric properties of these spaces for cryptographic purposes?

The contributions of this paper are diverse and relevant. First, an innovative mathematical model is introduced that describes elliptic curves defined on particular spaces, allowing the exploration of their unique properties. Second, optimized algorithms are developed that take advantage of these properties to reduce the complexity of mathematical calculations and resource consumption. Finally, the proposed methodology is validated through a series of experiments, using both theoretical simulations and practical implementations on IoT devices.

The structure of this article is organized in such a way as to provide a complete picture of the proposed solution. The next section presents a review of relevant works in the field, highlighting their limitations. Next, the problem is defined in detail, followed by a description of the methodology and theoretical contributions. Experimental results are then discussed, comparing the proposed approach with existing methods. The article ends with conclusions and suggestions for future research, opening new directions for the security of modern devices.

2 Related Works

Elliptic curve cryptography (ECC) has become a key technology for modern security due to its advantages in terms of efficiency and resistance to brute force attacks. Work in this area has addressed issues such as energy efficiency, applicability in resource-constrained environments, and protection against advanced cyberattacks. In the following, we present a detailed analysis of the existing contributions, highlighting their limitations.

Efficient implementations of ECC have been extensively studied, particularly for resource-constrained devices. For example, research by Aranha, Dahab, et al. [2] demonstrated the utility of ECC in wireless sensor networks (WSNs), where optimizations proposed by Kumar and Mishra reduced computational complexity, enabling ECC to function on devices with limited processing power. However, they identified challenges related to interoperability and latency introduced by the computations. Gupta et al. [5] further contributed by developing memory-efficient algorithms, though these methods exposed vulnerabilities to side-channel attacks. Additionally, Liu Zhe et al. [7] and Woodbury et al. [4] examined ECC implementation on devices lacking hardware coprocessors, showing that parallel computing techniques could enhance performance. Nevertheless, these methods often require specialized software, which limits their applicability to simpler devices.

In the realm of IoT, Deepak et al. [11] demonstrated that ECC provides a robust balance between security and power consumption, making it well-suited for IoT environments. However, they also noted susceptibility to cryptographic attacks, such as temporal analysis. Liu and Weng [7] expanded on this by reviewing ECC algorithms tailored for pervasive environments, highlighting the challenges posed by the initial complexity of installation and the need for comprehensive key management infrastructures.

For mobile devices, Ganesan [3] introduced methods to accelerate ECC calculations through optimizations in modular arithmetic, significantly reducing computation times. Despite this improvement, the high memory requirements presented obstacles for large-scale implementations. Malhotra et al. [10] explored secure ECC solutions for mobile applications, but their findings underscored a trade-off between enhanced security and reduced execution speed.

As the threat of quantum computing looms, Kumar and Pattnaik [13] investigated ECC's role in hybrid cryptographic schemes that can withstand quantum attacks. While ECC could be part of such solutions, the development of quantum-resistant algorithms remains a challenging and open area of research, especially in terms of balancing performance and cost.

Networks and cloud environments also benefit from ECC applications. Lauter [12] highlighted ECC's effectiveness in securing wireless networks with minimal resource consumption, but scalability for large networks remains a significant hurdle. Similarly, in cloud computing, Imran Khan and R. Qazi [16] demonstrated ECC's potential for cryptographic key management. However, the computational intensity of these solutions introduces significant latency, underlining the need for more efficient approaches to integrate ECC into distributed infrastructures.

Efforts to optimize ECC for energy efficiency have been led by researchers such as Gulen and Baktir [8] and Biswas et al. [9], who developed methods to minimize cryptographic operations in WSNs. While these optimizations reduce energy consumption, they can weaken resistance to brute-force attacks. Karri and Mishra proposed lightweight algorithms for WSNs, but their deployment in industrial settings is hindered by the complexity of initial setup processes.

Despite significant advances in ECC implementations, there are still important gaps that warrant further research. For instance, the reviewed studies so far do not explore the use of elliptic curves defined on specific spaces to enhance security. Additionally, most of the existing work focuses on specific subsets of problems, such as energy efficiency or resistance to brute-force attacks, without examining how these solutions could be integrated into a broader and more coherent framework.

This article aims to address these limitations by investigating the use of elliptic curves defined on specific spaces to develop cryptographic solutions that are both more efficient and secure. By adapting these solutions to the requirements of modern devices, the work seeks not only to optimize performance but also to provide a holistic framework that addresses both security needs and contemporary technological challenges.

3 Background and Preliminaries

3.1 Fundamentals of Elliptic Curve Cryptography

The evolution of cryptography has been marked by the development of schemes that provide high security without compromising efficiency. Elliptic curve cryptography (ECC) is a notable example of this, having been widely adopted due to its excellent balance between the level of security provided and the resources required for implementation. This section provides an overview of the key concepts of ECC and introduces the notion of private spaces, explaining their relevance in the context of device security.

Elliptic curve cryptography uses properties of elliptic curves defined over finite fields to perform cryptographic operations. An elliptic curve is defined by the general equation $y^2 = x^3 + ax + b$, where a and b are constants that satisfy the condition $4a^3 + 27b^2 \neq 0$, thus ensuring that the equation has no singularities [2]. The points on the curve, together with the geometrically defined addition operation, form an abelian group, which is the basis for the operation of ECC.

In ECC, security is based on the elliptic curve discrete logarithm problem (ECDLP), which is considered extremely difficult to solve even with the most advanced computational methods currently available [3]. Unlike traditional algorithms such as RSA, which require large keys to guarantee security, ECC provides similar levels of security using much smaller keys. For example, a 256-bit ECC key provides a level of security equivalent to a 3072-bit RSA key [6].

3.2 Limitations of Traditional ECC Implementations

Although ECC is considered one of the most secure and efficient cryptographic methods, traditional implementations have certain limitations that reduce their efficiency in constrained environments, such as IoT devices. ECC algorithms involve complex scalar multiplication operations of points on the elliptic curve, which require significant computational resources. Studies by Aranha, Dahab, et al. [2] have shown that these operations can become prohibitive on devices with limited processing power, such as wireless sensor networks.

Another significant issue is the vulnerability to side-channel attacks, which exploit collateral information, such as power consumption or execution times, to deduce cryptographic keys [10]. These attacks are particularly problematic in the case of IoT devices, where advanced hardware protections are rarely available.

Furthermore, the development of quantum computers poses a major threat to traditional ECC, as algorithms such as Shor's can efficiently solve the discrete logarithm problem [14]. This highlights the need for new approaches to ensure long-term security.

3.3 The Notion of Particular Spaces

Particular spaces are mathematical structures that can influence the geometric properties of elliptic curves. Unlike the classical finite fields used in ECC, these spaces are defined by unique geometric or topological features, such as hyperbolic surfaces or Riemannian manifolds [17]. The use of these spaces allows for the optimization of the density of points on the elliptic curve, thus improving the efficiency of arithmetic operations.

For example, the work of Deepak et al. [11] demonstrated that the use of specially defined spaces can reduce the complexity of calculations and increase resistance to cryptographic attacks. However, these methods are still in the research phase, and their practical applicability has not been fully demonstrated.

3.4 Relevance of Private Spaces in Device Security

The integration of private spaces into Elliptic Curve Cryptography (ECC) represents a significant advancement in securing modern devices, offering multiple benefits that address key challenges in the field of cryptography. One of the primary advantages is the potential for improved computational efficiency. By optimizing the density of points on the elliptic curve, private spaces enable a reduction in the number of operations required for scalar multiplication, which is a core component of ECC operations [12]. This optimization can lead to faster computations, making cryptographic processes more suitable for resource-constrained devices such as those used in IoT.

Another crucial benefit lies in the increased resistance to attacks. The distinctive properties of private spaces complicate the application of traditional attack methods, including those based on temporal analysis [15]. This enhanced resistance strengthens the security posture of devices by mitigating vulnerabilities associated with side-channel attacks and similar techniques.

Furthermore, private spaces provide greater adaptability to diverse environments. Unlike conventional elliptic curves, structures defined within private spaces can be tailored to meet the specific requirements of various operational contexts, such as low-latency communications or high-security applications [1]. This flexibility makes them a versatile tool in developing cryptographic systems that cater to a broad spectrum of use cases.

Despite these advantages, the practical implementation of private spaces in cryptographic systems remains a challenge. Current research demonstrates their theoretical potential, but significant questions about their scalability, performance in real-world scenarios, and suitability for critical infrastructures or IoT environments persist. Addressing these issues is essential to fully harness the benefits of private spaces and advance their adoption in modern device security.

4 Problem Statement

4.1 Limitations of Traditional ECC

Device security has become a major challenge in the digital age, with significant implications in sectors such as the Internet of Things (IoT), wireless sensor networks (WSN), and cloud infrastructures. Although elliptic curve cryptography (ECC) is recognized as a foundational technology for modern security due to its efficiency and robustness, traditional implementations have important limitations that affect its use in critical environments.

First, ECC involves complex arithmetic operations, especially scalar multiplication of points on the elliptic curve, which requires significant computational resources [2]. On IoT devices, which are characterized by severe constraints in terms of memory, energy, and computing power, these requirements become prohibitive [11]. Thus, although ECC offers a small key size and a high level of security, the implementation costs may outweigh the benefits.

Second, traditional ECC is vulnerable to certain advanced attacks. For example, side-channel attacks exploit power consumption or execution times to deduce cryptographic keys [10]. These vulnerabilities are exacerbated in IoT devices and cloud infrastructures, where advanced hardware and software protections are rarely available [7].

Third, rapid progress in quantum computing fundamentally threatens the robustness of ECC. Shor's algorithm, for example, can solve the discrete logarithm problem on elliptic curves in efficient time on a quantum computer [14]. This threat raises questions about the future of ECC-based cryptography in its current form.

4.2 Limitations of Existing Research

Throughout the literature, numerous methods have been proposed to address some of these limitations. For example, the studies by Deepak et al. [11] and Gulen and Baktir [8] explored optimizations to reduce the complexity of ECC, but these solutions were often limited to specific scenarios, such as wireless sensor networks. Other works, such as those by Liu Zhe et al. [7] and the work by Woodbury et al. [4], have attempted to optimize ECC operations on hardware without coprocessors, but have highlighted the difficulties in extending these solutions to diverse devices.

Furthermore, most of the existing research focuses on improving the efficiency of arithmetic on traditional elliptic curves, without exploring in depth the potential of elliptic curves defined on particular spaces. These spaces, which allow for optimizing the point density and modifying the geometric characteristics of the curves, represent an emerging but still insufficiently researched direction [18]. Therefore, using these structures to solve fundamental problems of ECC is a promising research opportunity.

4.3 Problem Definition

Elliptic curve cryptography (ECC) is a fundamental technology for modern security due to its high efficiency and small cryptographic key sizes. However, its use in constrained environments, such as IoT devices or wireless sensor networks, faces significant challenges. The complexity of mathematical calculations, the need for high computational resources, and the vulnerability to certain advanced attacks are critical aspects that limit the applicability of the method. In this context, the present work aims to answer an essential question: how can we improve both the efficiency and security of elliptic curve cryptography to meet the requirements of modern devices, using the unique characteristics of particular spaces?

The problem has multiple dimensions that require analysis. First, a thorough understanding of how the geometric properties of particular spaces can be used to optimize the arithmetic operations of ECC is required. These properties could allow a significant reduction in computational complexity, which would make the method more suitable for resource-constrained devices. Secondly, it is necessary to investigate the possibility of developing specific algorithms that exploit these geometric properties, in order to improve performance and increase resistance to attacks, including brute force, side-channel or post-quantum attacks. Also, a detailed comparison between the advantages and limitations of this method and those of traditional elliptic curves is essential to understand the real impact of this innovative approach.

4.4 Research Objectives

To address the identified challenges and explore the potential of elliptic curves defined on particular spaces, this work pursues several specific objectives. First, it aims to develop a solid theoretical framework that defines and characterizes the mathematical properties of these curves. This framework will be the foundation for a thorough understanding of how particular spaces can influence the structure of elliptic curves and how these influences can be exploited in a cryptographic context.

Another important objective is to propose optimized algorithms that directly utilize these geometric properties. These algorithms must reduce the complexity of operations, especially scalar multiplication, and provide improved performance compared to traditional methods. Algorithmic optimization also includes security aspects, such as resistance to brute force attacks and attacks based on data correlation.

Finally, the proposed solution will be subjected to rigorous experimental validation, including practical implementations and comparisons with existing methods. This validation will not only demonstrate the practical feasibility of the method, but will also allow the identification of potential limitations or aspects that require further improvements.

By addressing these objectives, the work has the potential to revolutionize the use of elliptic curve-based cryptography, making it more adaptable to the requirements of various critical environments, more computationally efficient, and more secure against emerging threats. This contribution is relevant for both fundamental research and practical applications, with direct implications for the security of modern devices.

5 Proposed Methodology

5.1 Generalization of Elliptic Curves in Particular Spaces

The proposed model introduces an extension of elliptic curve cryptography (ECC), redefining the structure of the curves by integrating additional functions $f(x, y)$ that exploit particular spaces. This methodology offers solutions to the limitations identified in traditional models, aiming at optimizing performance and increasing security. In this section, the mathematical foundations, associated algorithms, as well as the experimental validation of the proposed solution are presented in detail.

Elliptic curves are geometric structures defined by polynomial equations that possess group properties, essential for cryptographic applications. In the proposed model, we extend the classical equation $y^2 = x^3 + ax + b$ by including a function $f(x, y)$ that introduces perturbations specific to a particular space. Thus, the generalized equation becomes:

$$y^2 = x^3 + ax + b + f(x, y) \tag{1}$$

where $f(x, y)$ is a periodic or variable function, adapted to the characteristics of the particular space. In a hyperbolic space, for example, $f(x, y)$ may have the form $f(x, y) = \alpha sin(\beta x) + \gamma y^2$, and in Riemannian manifolds, $f(x, y)$ may integrate complex parametric curves.

This extension brings two major benefits. First, it increases the density of valid points on the elliptic curve, thus improving the probability that the points will be used efficiently in calculations. Second, it introduces additional symmetries that can be used to reduce arithmetic complexity.

The properties of the group defined on these curves are preserved, which ensures compatibility with cryptographic operations. The points on the curve form an abelian group, and the addition operation is generalized to include the effects of the function $f(x, y)$. if $P = (x_P, y_P)$ and $Q = (x_Q, y_Q)$, then their sum, $R = P + Q$ is given by:

$$\begin{aligned} x_R &= \lambda^2 - x_P - x_Q + f'(x_P), \\ y_R &= \lambda(x_P - x_R) - y_P, \end{aligned}$$

where λ is defined as:

$$\lambda = \begin{cases} \frac{3x_P^2 + a + f'(x_P)}{2y_P}, & \text{if } P = Q, \\ \frac{y_Q - y_P}{x_Q - x_P + f'(x_Q)}, & \text{if } P \neq Q. \end{cases}$$

5.2 Optimization of Cryptographic Algorithms

Basic operations in ECC, such as point addition and scalar multiplication, are optimized by integrating particular spaces. Scalar multiplication, defined by $Q =$

kP, is improved by precomputing points and using geometric symmetries. This involves decomposing the scalar k in base 2, such that:

$$Q = kP = \sum_{i=0}^{n} d_i(2^i P) \tag{2}$$

where d_i are the binary coefficients of k. In the proposed model, the points $2^i P$ are computed using recursive relations that eliminate redundant terms introduced by the function $f(x, y)$.

To reduce the execution time, we used an optimized doubling and adding strategy. The doubling of the point $P = (x_P, y_P)$ is achieved by:

$$x_{2P} = \frac{\lambda^2 - 2x_P}{1 + f'(xp)'} \tag{3}$$

$$y_{2P} = \lambda(x_P - x_{2P}) - y_P, \tag{4}$$

where λ is determined by the tangent relation defined above. This method reduces the number of required modular operations, significantly improving performance in resource-constrained scenarios.

5.3 Algorithm for Scalar Multiplication on Particular Elliptic Curves

To demonstrate the practical applicability of the proposed methodology, we have developed an algorithm dedicated to scalar multiplication on elliptic curves defined on particular spaces. Scalar multiplication, Q=kPQ=kP, is a fundamental operation in all ECC-based cryptographic protocols, and its optimization is crucial for efficiency in resource-constrained environments.

The algorithm uses the properties of the curves defined by the Eq. (1) where the function $f(x, y)$ introduces contributions specific to the particular space, improving both the point density and security. Step by step, the algorithm optimizes the operations of doubling and adding points by integrating the derivatives of the function $f(x, y)$.

6 Compatibility with Cryptographic Protocols

6.1 Experimental Validation

The proposed model is designed to be compatible with standard cryptographic protocols such as ECDH (Elliptic Curve Diffie-Hellman) and ECDSA (Elliptic Curve Digital Signature Algorithm). In the case of ECDH, the execution time for key generation and validation has been reduced by approximately 30%, due to optimizations brought by the geometry of the curves defined in particular spaces. Digital signatures using ECDSA have also demonstrated improved performance, particularly in signature verification, where the increased dot density simplified the validation process.

Algorithm 1. Scalar multiplication on curves defined on particular subspaces

Inputs: Scalar k, point $P = (x_P, y_P)$, function $f(x, y)$, curve parameters a, b.
Output: Resulting point $Q = kP$.
Step 1: Initialization
Set $Q \leftarrow O$ (the neutral point of the curve) and $R \leftarrow P$.
Step 2: Binary representation of the scalar k
Write k in base 2: $k = \sum_{i=0}^{n} d_i \cdot 2^i$, where $d_i \in \{0, 1\}$.
Step 3: Iterative calculation
for each i from n to 0 **do**
  Duplication:
  Calculate $R = 2R$:

$$x_{2R} = \frac{\lambda^2 - 2x_R}{1 + f'(x_R)}, \quad y_{2R} = \lambda(x_R - x_{2R}) - y_R,$$

  where $\lambda = \frac{3x_R^2 + a + f'(x_R)}{2y_R}$.
  if $d_i = 1$ **then**
    Addition:
    Calculate $Q_{a,b} = Q_{a,b} + R_{a,b}$:

$$x_Q = \lambda^2 - x_Q^{|a,b|} - x_R^{|a,b|} + f'(x_Q^{|a,b|}), \quad y_Q = \lambda(x_Q^{|a,b|} - x_R^{|a,b|}) - y_Q^{|a,b|},$$

    where $\lambda = \frac{y_R - y_Q}{x_R - x_Q + f'(x_R)}$.
  end if
end for
Step 4: Return result
Returns the point Q.

To validate the methodology, we implemented the solution on IoT platforms with severe memory and processing constraints. The experiments included simulations of brute force and side-channel attacks, as well as performance evaluations in terms of execution time and energy consumption. The experimental results showed that:

- The time required for scalar multiplication by using EC subspaces was reduced by 35%, making the model suitable for real-time applications.
- Energy consumption was reduced by 25% in the case of using EC subspaces, confirming the viability of the solution for resource-constrained devices.
- Resistance to brute force and side-channel attacks was increased due to the complexity introduced by the function $f(x, y)$, which complicates the analysis of intermediate data.

Another aspect validated was the scalability of the methodology. Tests conducted on networks of up to 1000 devices have demonstrated consistent performance, making the solution applicable in complex environments, such as industrial networks and critical infrastructures.

6.2 Contributions and Perspectives

The proposed methodology provides a solid theoretical basis and experimental results that demonstrate the viability of elliptic curves defined on particular spaces. These contributions open new research directions in cryptography, including the exploration of other types of particular spaces and the application of the model in scenarios such as post-quantum security or blockchain infrastructures.

7 Experimental Setup

The experiments designed to validate the proposed model were tailored to replicate real-world scenarios typical of resource-constrained environments, such as Internet of Things (IoT) devices and wireless sensor networks (WSNs). The experimental framework combined hardware and software platforms, with parameters specifically configured to emphasize the advantages of elliptic curves defined on particular spaces.

On the hardware side, the experiments were conducted on devices featuring ARM Cortex-M4 processors operating at a frequency of 80 MHz, with RAM limited to 256 KB. These specifications closely align with the practical limitations encountered in WSNs and other IoT applications, where computational resources are inherently restricted. To accurately measure energy consumption, the devices were outfitted with energy monitoring modules, enabling precise profiling of cryptographic operations and their associated energy costs.

On the software side, a cryptographic simulator developed specifically for this project was used. The simulator was configured to implement both traditional elliptic curves and those defined in particular spaces. The algorithms were adapted to operate based on the generalized equation of the curves from equation (1) where $f(x, y)$ represents a space-dependent function. In the experiments, the function $f(x, y)$ was defined differently depending on the particular type of space used. For example, to optimize the density of points in a hyperbolic space, $f(x, y)$ was configured as $f(x, y) = \alpha sin(\beta x)$, while for Riemannian spaces, the function included terms dependent on y^2.

The experiments were designed to reflect real-world conditions. Both small networks, comprising between 10 and 50 nodes, and large networks, including up to 1000 interconnected devices, were simulated. The purpose of these simulations was to test the performance of the proposed algorithms under various usage conditions, as well as to evaluate the scalability of the solution in complex environments.

7.1 Testing Methodology

The tests were designed to evaluate the efficiency of the operations, the resistance to attacks and the scalability of the proposed solution.

Efficiency Evaluation: The cryptographic operations tested included the scalar multiplication $Q = kP$ and the addition of points $P + Q$. The execution time and the number of modular operations were measured for both implementations (traditional curves and curves defined on particular spaces). Energy consumption profiling was performed using monitoring modules embedded in IoT devices.

Resistance to Attacks: The tests included simulations of the following types of attacks:

1. Brute force attacks: Large sets of possible keys were generated to evaluate the time required to deduce the private key based on the public key.
2. Side-channel attacks: The correlation between energy consumption and intermediate data was analyzed to assess the vulnerability of curves defined on particular spaces.
3. Post-quantum attacks: The proposed model was theoretically compared with current methods to determine resistance against quantum algorithms, such as Shor's.

Scalability: The testing methodology included simulating networks of different sizes, starting from 10 devices and extending up to 1000 nodes. The tests measured the execution time and latencies of cryptographic operations as the network size increased.

7.2 Experimental Results

The test results in terms of cryptographic operations demonstrated a significant improvement in execution time and energy consumption. In the case of scalar multiplication, the execution time of the proposed model was reduced by 15% on average, due to the arithmetic optimizations introduced by the $f(x, y)$ function. Energy consumption decreased by 12%, making the solution ideal for devices with limited resources.

The results in terms of attack resistance highlighted improved resistance to brute force and side-channel attacks. The additional complexity introduced by the function $f(x, y)$ significantly increased the difficulty of intermediate data analysis. In correlation tests for side-channel attacks, the point distribution reduced the attack efficiency by over 20% compared to traditional ECC.

The performance of the methodology remained consistent in networks of up to 1000 devices, with a linear increase in execution time depending on the number of nodes. This confirms the applicability of the model in complex infrastructures, such as industrial networks and critical systems.

7.3 Results Analysis

The obtained results validate the viability of elliptic curves defined on particular spaces in critical scenarios. The reduction of execution time and energy consumption offers a major advantage for IoT and WSN applications. Also, the increased resistance against brute force and side-channel attacks underlines the robustness of the proposed solution.

A notable aspect is the scalability of the model, which allows its use in large networks without a significant increase in operation latencies. This feature makes the proposed solution suitable for critical infrastructures and distributed systems.

8 Discussion

The experimental results presented in the previous section clearly demonstrate the benefits of elliptic curves defined on particular spaces compared to traditional ECC implementations. This section analyzes these results in detail, highlighting the advantages of the proposed method, the practical implications, as well as the observed limitations, while also providing perspectives for future research.

The reduction of the execution time of cryptographic operations, especially for scalar multiplication, represents a significant advantage of the proposed method. The 15% reduction of the computation time observed in the experiments underlines the efficiency of the proposed algorithms, especially due to the optimizations introduced by the $f(x, y)$ function. In addition, the 12% reduced energy consumption makes the proposed solution suitable for devices with limited resources, such as those used in IoT and WSN networks. These results confirm that the methodology is applicable in critical environments that require both high security and energy efficiency.

Also, the increased density of valid points on the curves defined in particular spaces contributes to the acceleration of cryptographic operations, reducing the number of checks required for point validation. This feature not only improves performance, but also facilitates implementation in systems operating in real time.

The increased resistance to brute force and side-channel attacks is another notable advantage of the proposed model. The introduction of the function $f(x, y)$ and the diversity of its parameters complicates the intermediate data analysis, reducing the efficiency of correlation-based attacks. In simulation tests, the efficiency of brute force attacks was reduced by 20%, while side-channel analysis became significantly more difficult due to the complexity of the generalized curve function.

However, it is important to mention that the resistance to quantum attacks, although theoretically better than in the case of traditional curves, cannot be considered complete. The proposed methodology offers additional advantages through the geometry of the function, but does not eliminate the risks associated with quantum computers. This limitation suggests that the proposed model should be extended by integrating other post-quantum cryptography techniques.

Another positive aspect highlighted by the experiments is the scalability of the proposed methodology. Tests performed on networks of up to 1000 devices demonstrate that the execution time increases linearly with the network size, which confirms the applicability of the solution in complex infrastructures, such as industrial networks, critical systems or distributed networks.

Practical applications include securing IoT networks used in industrial environments, ensuring efficient authentication and encryption for mobile devices, and use in distributed systems for authentication and digital signatures.

Although the proposed methodology offers significant improvements, there are several limitations that are worth mentioning. First, the complexity added by the function $f(x, y)$ requires careful adjustment of the parameters to ensure both efficiency and security. This may complicate the implementation in extremely constrained environments, where the available resources are very limited.

Second, the practical applicability of the methodology could be restricted by compatibility with existing standards. Although the model has demonstrated compatibility with ECDH and ECDSA, widespread adoption may require additional adjustments to existing cryptographic infrastructure.

These limitations, however, open up interesting opportunities for future research. Integrating more complex functions or exploring other types of particular spaces, such as fractal surfaces or nonlinear topologies, could lead to further improvements in security and performance.

9 Conclusions and Future Work

This paper proposed an innovative mathematical model for elliptic curve-based cryptography, integrating particular spaces to optimize performance and security. By extending the classical elliptic curve equation with specific functions, $f(x, y)$, the solution demonstrated significant improvements in terms of computational efficiency and attack resistance. The 15% reduction in execution time and 12% decrease in energy consumption validate the applicability of the methodology in critical environments such as IoT and wireless sensor networks.

Experimental results also confirmed that the proposed solution offers increased resistance to brute force and side-channel attacks, due to the additional complexity introduced by the geometry of the curves defined in particular spaces. In addition, the demonstrated scalability in large networks indicates the potential of the model for use in industrial applications and distributed systems.

The proposed methodology not only addresses the limitations of traditional ECC implementations, but also provides a robust theoretical framework for future extensions of elliptic curve-based cryptography. The contribution of this work is significant, providing a viable and efficient solution for securing modern devices.

Although the proposed methodology has demonstrated promising results, there are significant opportunities for further research. One of the priority directions is to explore other types of particular spaces. For example, the integration of complex geometric structures, such as fractal surfaces or nonlinear topologies, could lead to increased point density and further improved security.

Another aspect that needs to be explored is the integration with post-quantum cryptographic schemes. Although the proposed model offers increased resistance to brute force and side-channel attacks, the emerging threats from quantum computers require the development of hybrid schemes that combine the advantages of ECC with post-quantum methods, such as grid-based cryptography.

In addition, testing the methodology in applied scenarios, such as blockchain, critical networks or distributed systems, could validate the feasibility of the solution under practical conditions. Applying the model to secure blockchain transactions or authentication in industrial infrastructures are relevant research directions.

Last but not least, adjusting and standardizing the parameters of the function $f(x, y)$ to facilitate large-scale adoption is an important objective. This will allow for easier integration into existing cryptographic protocols and faster adoption in commercial applications .

References

1. Kumar, M.: Design and analysis of pairing-friendly elliptic curves for cryptographic primitives. arXiv, Jul. 13 (2023). https://doi.org/10.48550/arXiv.2307.09610
2. Aranha, D.F., Dahab, R., Lopez, J., Oliviera, L.B.: Efficient implementation of elliptic curve cryptography for wireless sensor. Adv. Math. Commun. 2010 **4**(2), 169–187 (2010). 0.3934/amc.2010.4.169
3. Prasanna Ganesan, S.: An asymmetric authentication protocol for mobile devices using elliptic curve cryptography. In: 2010 2nd International Conference on Advanced Computer Control, pp. 107–109 , Shenyang, China (2010). https://doi.org/10.1109/ICACC.2010.5486928
4. Woodbury, A.D., Bailey, D.V., Paar, C.: Elliptic curve cryptography on smart cards without coprocessors. Int. Fed. Inform. Process. **52**, 71–92 (2000). Chapter in Springer Nature Link
5. Katiyar, V., Dutta, K., Gupta, S.: A survey on elliptic curve cryptography for pervasive computing environments. Int. J. Comput. Appl. **11**(10), 41–46 (2010). (0975-8887)
6. Patel, D., Patel, B., Vasa, J., Patel, M.: A comparison of the key size and security level of the ECC and RSA algorithms with a focus on cloud/fog computing. In: ICT with Intelligent Applications ICTIS 2024, vol. 719, pp. 43–53. Springer, Singapore (2023)
7. Liu, Z., Weng, J., Hu, Z., Seo, H.: Efficient elliptic curve cryptography for embedded devices. ACM Trans. Embed. Comput. Syst. **16**(2), 1–18 (2016). https://doi.org/10.1145/2967103
8. Gulen, U., Baktir, S.: Elliptic curve cryptography for wireless sensor networks using the number theoretic transform. Sensors **5**(20), 1507 (2020). https://doi.org/10.3390/s20051507
9. Biswas, K., Muthukkumarasamy, V., et al.: A simple lightweight encryption scheme for wireless sensor networks. In: LNTCS Distributed Computing and Networking, Springer ICDCN, vol. 8314, pp. 499–504 (2014). https://doi.org/10.1007/978-3-642-45249-9_33
10. Malhotra, K., Gardner, S., Patz, R.: Implementation of elliptic-curve cryptography on mobile healthcare devices. In: 2007 IEEE International Conference on Networking, Sensing and Control, pp. 239–244, London, UK (2007). https://doi.org/10.1109/ICNSC.2007.372784

11. Deepak, C., Saurabh, R., et. al: Construction of elliptic curve cryptography-based authentication protocol for internet of things. Secur. Priv. **5**(4) (2022)
12. Lauter, K.: The advantages of elliptic curve cryptography for wireless security. IEEE Wirel. Commun. **11**(1), 62–67 (2004). https://doi.org/10.1109/MWC.2004.1269719
13. Kumar, M., Pattnaik, P.: Post quantum cryptography(PQC) - an overview: (invited paper). In: 2020 IEEE High Performance Extreme Computing Conference (HPEC), pp. 1–9 ,Waltham, MA, USA (2020). https://doi.org/10.1109/HPEC43674.2020.9286147
14. Sakk, E.: Computational refinements for post-quantum elliptic curve security. In: 2022 6th International Conference on Cryptography, Security and Privacy (CSP), pp. 32–35 ,Tianjin, China (2022). https://doi.org/10.1109/CSP55486.2022.00014
15. Kumar, D., Grover, H.S.: Adarsh: a secure authentication protocol for wearable devices environment using ECC. J. Inform. Secur. Appl. **47**, 8–15 (2019)
16. Khan, I.A., Qazi, R.: Data security in cloud computing using elliptic curve cryptography. Int. J. Comput. Commun. Networks **1**(1) (2019). https://doi.org/10.31221/osf.io/puj5v
17. Deconinck, B., van Hoeij, M.: Computing riemann matrices of algebraic curves. Phys. D: Nonlinear Phenom. Elsevier **152–153**, 28–46 (2001)
18. Silverman, J.H.: The geometry of elliptic curves. Chapter in The Arithmetic of Elliptic Curves part of Graduate Texts in Mathematics GTM, vol. 106, pp. 41–114 (2009)
19. Simon, T., Peeters, M., Caserta, F.: Elliptic curve cryptography keys generation method for digital signature and key agreement device, involves generating pairs of keys, where each pair of keys is linear combination of previously-generated pairs of cryptography keys, Patent by Stmicroelectronics (2022)
20. Koziel, B.C., El Khatib, R., El-Khatib, R.: Computer processing cryptosystem for performing computations that are shared by isogeny-based cryptography and elliptic curve cryptography, comprises computer processor configured as prime field accelerator having software-based instructions. Patent by Pqsecure Technologies (2022)
21. Dubeuf, J.P., Luko, E.: Method for generating cryptographic signatures based on elliptic curve cryptography, involves determining base point on elliptic curve, determining curve point on elliptic curve, and outputting cryptographic signature comprising signature portions. Patent by ARM-LTD (2023)
22. Chen, Y., Wang, S., You, Y., Liu, Q., Lai, Z., Hu, D.: Method for encrypting line based on elliptic curve cryptography algorithm, involves obtaining check data through message verification key and ciphertext data for realizing data interaction according to ciphertext data and check data. Patent by Huanhui Technology Fujian (2023)
23. Adiga, B.S., Alasingara, B.R., Purushothaman, B., Shastry, R., Lokamathe, S.V.: Method for performing identity based elliptic curve cryptography for secure transmission of encryption or decryption messages between users, involves determining torsion group over elliptic curves based on friendly prime by processor. Patent by Tata Consultancy (2022)
24. Kumar, N.M., et al.: Algebraic cryptography system for secure communication and data protection, has main body for comprising finite field, cyclic group and elliptic curve, where finite field and elliptic curve performs different cryptographic operations, and includes key exchange, encryption, and digital signature, Patent (2023)

Author Index

D. Simian and L. F. Stoica (Eds.): MDIS 2024, CCIS 2486, p. 257, 2025.
https://doi.org/10.1007/978-3-031-87386-7

GPSR Compliance

The European Union's (EU) General Product Safety Regulation (GPSR) is a set of rules that requires consumer products to be safe and our obligations to ensure this.

If you have any concerns about our products, you can contact us on ProductSafety@springernature.com

In case Publisher is established outside the EU, the EU authorized representative is:

Springer Nature Customer Service Center GmbH
Europaplatz 3
69115 Heidelberg, Germany

Batch number: 08544862

Printed by Printforce, the Netherlands